Barolo *Library*

餐 桌 上 的
義 大 利 酒 王

我 的 100 瓶 巴 洛 羅 紅 酒 評 審 指 南

2015 年 1 月

　　正式開始「Barolo Library 餐桌上的義大利酒王」計劃

2015 年 4 月

　　100 瓶新上市的巴洛羅紅酒於義大利收集完畢並集裝出口

2015 年 6 月

　　「雲朗酒窖」盲飲活動正式於亞州登場

2015 年 7 月

　　「Barolo Library 餐桌上的義大利酒王」此書正式出版

待續 ...

January 2015

The project "Barolo Library" was launched.

April 2015

100 different new-vintage Barolo were gathered. Collection completes.

June 2015

The opening of Barolo Library in Taipei, Taiwan.

July 2015

The book "Barolo Library" is published.

To be continue...

目 錄
Contents

前 言
Preface

評 審 會
Judge Tastings

索　引
Index

參與 Barolo Library 的各酒莊負責人合影
Owners of different Barolo wineries who participate in 《Barolo Library》

各酒莊負責人 owners of different Barolo winery

巴洛維協會理事長 Federico Scazello

「時間」

在義大利、似乎停留在某一時期

你能夠想像、2015 年還要透過傳真機才能買酒嗎？

2014 年 12 月，我第一次向「巴洛羅紅酒官方協會」理事長法德里克‧斯卡柴諾與「北義帕拉維嘉葡萄酒協會」理事長瑪雀拉‧碧揚卡兩位北義大利葡萄酒領導人物、討論「Barolo Library 餐桌上的義大利酒王」的這個構想；這是一個「把每一年在北義整個村莊的葡萄酒新酒上市傳統（在義大利文稱為 Il Barbaresco di Tavola）、一整個搬到台灣進行並且出書」的計畫。猶稀記得，當我把整個計畫緩緩告訴這兩位年輕代的酒界領導人時，雖然他們的眼瞳閃爍著激動的光芒，其表情卻無比平靜地告訴我，我即將面臨的最大問題 —— 義大利巴洛羅酒莊的傳統生態。

「時間」在義大利、似乎停留在某一時期，一直到今天，部分酒莊依舊維持著 80 年代的經營模式，那沒有網路只有傳真機、沒有手機只有舊式撥號電話、沒有銷售經銷網只有家族產業和 30 年老客戶的直接登門購買。許多巴洛羅百年酒莊的經營模式和態度，充滿了每個人不同的臉、笑容、文字和人情味；這、不是一個講求數字的世界；產量有限，亦無法講究數字。義大利巴洛羅紅酒之所以稱為「酒中之王、王者之酒」，其重要性可以從其歷史、葡萄品種、陳釀手法與其橡木桶、傳統與新潮之爭、年份及其層次表現、11 個法定村莊的不同特色、各葡萄坡的鈣含量、各酒莊引以為傲的日照方向與風向、以及各酒莊釀酒師的陳釀哲學等來談。關於巴洛羅紅酒一切的一切，正是我情不自禁愛上「他」及其文化的千萬個理由。

在寫這本書的時候，很多人問我為什麼要寫這本書，怎麼會有這樣的想法。我笑而不語，因為文字很難形容當下我的感受；在大量品飲並從超過兩百家巴洛羅酒莊中、選擇表現出眾的葡萄坡與其葡萄酒、進而驚訝於自己的認知依舊貧脊，而這正是希臘哲學家所謂的「知道的越多，我越了解我什麼都不知道；The more I know, the more I understand I don't know」時；在與各家酒莊負責人或釀酒師談話而意外結交好友並相知相惜時；在一一拜訪各家酒莊並了解其家族歷史與釀酒哲學，並為其堅持與倔強所感動時；在邀請義大利米其林主廚、飯店行政主廚與伺酒師、酒評家及葡萄酒愛好者共同品評，評審會當天我享受那每一分每一秒的興奮與滿足時。準備這本書的過程，我得到的「快樂」已經無價。

感謝雲朗集團在台灣推動義大利文化不遺餘力；感謝夫人和執行長對晚輩的提攜厚愛；感謝我的父母願意放手讓我的夢想飛翔；感謝我在義大利的良師益友瑪雀拉·碧揚卡、馬立歐·婭吉歐翁和馬可·安東努齊；感謝多年照顧我並把我當家人對待的芭爾巴拉·索都、法布利奇歐·婭斯可利、以及數年來與我共享盤中美食、杯中美酒、提供友情與溫暖的歐洲朋友們。謝謝你們存在於我的人生，我會一直微笑珍惜！

關於作者

義大利美食教育家黃筱雯，政治大學雙學士，台灣大學碩士畢，領有多張義大利美食相關證照（品油專家、品酒師、巴薩米克醋鑑賞、松露鑑賞等），目前為 CLUBalogue 義大利美食教育中心創辦人。負責橄欖油、巴薩米克醋、松露等義大利食材教育推廣，邀請國際比賽冠軍達人與義大利米其林主廚來訪亞洲。此外，她於法義邊境經營由 17 世紀修道院改建的個人民宿，也在義大利開始釀造傳統型巴薩米克醋，是位懂美食、文化、教育的生活主義者。

關於義大利美食教育中心：www.CLUBalogue.com

Time

It remains unmoved in Italy

In the year of 2015, FAX is still in options for purchasing a bottle of Barolo...

I remember, in December 2014 when it was the first time I shared my idea of 《Barolo Library》 with two of my dear Italian friends, Marcella Bianco, President of Pelaverga Association and Federico Scazello, President of Enoteca di Barolo. I can still recall their calm faces with encouraging eyes, telling me situations that I could be facing. The project of Barolo Library is to bring "Il Barbaresco a Tavola" from Italy to Taipei which includes the purchase of 100 different Barolo of different wineries and of different crus (obviously each Barolo more than one bottle), inviting chefs and wine experts to conduct judge tastings in Italy, cooperation with Alessandro Masnaghetti (the Map Man), delivery of pallets of Barolo to the other side of the world (Asia), and finally, the publication of this book.

The biggest obstacle and the beauty of Barolo reality is that sometimes the concept of "time" seems to remain unmoved since the 80s: the history, the grapes, the families, the traditions of using big oak Barrels, the cru/exposition/soil composition/ vinification, the philosophy of enologists and the love that people share through a bottle of Barolo. Up to now, there are still some wineries remain how and who they were 30 years ago: no internet but only fax machine, no mobile phone but only dial phone, no distribution net but family business and door-to-door same clients for the past 30 years. I could say that there have been some wonderful and necessary revolution in the recent decade and surely most wineries use internet and email. Yet, the optimal value of Barolo being "King of Wine" and "Wine of King" lies on the fact that you can SEE and you can FEEL the love and passion of Italian people toward its tradition and its culture. And this is the reason why I also fall in love with Italian wine.

Many people asked me why I write this book and how I come up with this idea of putting together 100 Barolo wine and almost 50 experts in one concept. To cut the story short and to satisfy those doubts and curiosity, my answer is simple: I do all these work because of the priceless happiness it gives me every day.

Special thanks to LDC Hotel Group; to Ruth and Nilson; to my parents; to Marcella Bianco, Mario Adrion and Marco Antonucci; to Barbara Soddu, Fabrizio Ascheri, and to many others who have shared moment of your life with me. Thank you for being in my life and I will continue with the same smile despite of all difficulties it may be on the path.

Author

Founder of CLUBalogue; Oil Sommelier; Wine Sommelier (On progress); Master in law and politics of National Taiwan university; Double Bachelor in Philosophy and Political science in National Cheng Chi University; Proud to be Taiwanese while in her blood there's also Japanese. She is passionate about every minute of life and in deep love with Italian food and wine. She starts to be backpacker since age 16 and travels around the world ever since. She loves meeting people and looks forward to meeting you in tasting event. Other publication:
《10 Ingredients of 10 Traditional Italian Food Family》

Founder of Modern Italian Cuisine - Gualtiero Marchesi

義大利第一位三星米其林主廚
今年 86 歲的國寶級教父推薦本書

This is the thinking on Barolo that Gualtiero Marchesi writes for this book. Thank you, Gualtiero Marchesi. May the readers, though possibly on the other side of the world, have a chance to enjoy the greatness of Italian food and wine.

Xiaowen Huang

以下為義大利國寶級教父葛提耶洛‧瑪切希本人為此書所寫的推薦序。在此感謝年已 86 歲依舊活躍於義大利食品界的第一主廚分享他對巴洛羅紅酒的想法。謝謝你，葛提耶洛。願正在閱讀此書的你（們），雖然可能位處於世界的另一個角落，得一良機享受真正義大利飲食文化之美好。

本書作者
黃筱雯

Marchesi per Barolo Library（義大利原文）

Quando penso al Barolo, forse il più grande vino italiano, penso ad un vino da gustare come un cibo, corposo, strutturato, da bere "a bocconi".

Per questo, penso che non sia facilissimo creare degli abbinamenti con dei piatti che gli stiano alla pari.

Il Barolo è una delle tante magnifiche risorse della cultura enogastronomica italiana, fondata sulla diversità dei microclimi.

Gualtiero Marchesi

Marchesi for Barolo Library (English translation)

When I think of Barolo, possibly being the greatest wine of Italy, I consider to enjoy it as ONE DISH with full-body, good structure, and to drink "piece by piece".

For this reason (that Barolo is such great wine), I think it is not easy to create dishes of equal greatness for a proper matching.

The Barolo wine is one of the most wonderful and valuable resources of Italian food and wine culture based on diversity of microclimates./Gualtiero Marchesi

瑪切希推薦序（中文翻譯）

當我想著義大利巴洛羅紅酒時，好比品嚐一道絕世佳餚般，我一口一口細細思量「他」豐滿的酒體與完美的層次結構。美酒理應搭配美食，然我認為要能做出一道如同巴洛羅紅酒一般完美的菜餚與之搭配，絕非易事。正是因為巴洛羅地區特殊環境（微氣候）具多元性，「巴洛羅紅酒」這個名詞也成為了義大利飲食文化的最佳代表。

葛提耶洛‧瑪切希

巴洛羅的風土條件

　　對我來說、愛上研究巴洛羅紅酒的一大要素在於、每一瓶來自不同酒莊的巴洛羅紅酒與其「風土條件 (Terroir，註[1])」的關係是多元的、系統化的且幾近科學的；每品嘗一杯巴洛羅紅酒，腦海中同時編織聯想著那官方地圖上、十一個巴洛羅村落 (註[2]) 的所在地；每嘗一口，便在那地圖上註記一筆。當一個人累積品嘗一定數量、來自不同酒莊不同村落的巴洛羅紅酒時，當一個人能夠用味蕾刻劃出、理解到屬於自己的巴洛羅紅酒地圖時，亦能體會「土地」與「葡萄酒」的深度關聯。關於這部分的更多資訊，我推薦三個網站：

1. www.BaroloLibrary.com　2. www.EnotecadelBarolo.it　3. www.enogea.it

　　用「盲飲」的方式來了解巴洛羅紅酒世界中那「土地」與「葡萄酒」的關聯、是一大樂趣也是一大挑戰。由於這官方地圖上的 11 個村莊中、各葡萄坡土壤鈣含量與層次結構皆不同，其生產之巴洛羅紅酒風味亦不同。我們大致可分為西北半邊的女性巴洛羅紅酒 (Female Barolo) 與東南半邊的男性巴洛羅紅酒 (Male Barolo)。娜比歐羅葡萄 (Nebbiolo Grape，註[3]) 普遍具有明顯果實香韻 (Fruity) 與無可取代的美妙酸度 (Acidity)。然來自前者的尾韻果香擁有著如女性般的優雅飄逸，如櫻桃 (Cherry) 和黑醋栗 (Black Current) 等；來自後者的尾韻則擁有著如男性般的強勁力道、充分表現出橡木 (Wood) 與皮革 (Leather)。此為學習盲飲巴洛羅紅酒初學者的基本原則 (註[4])。

註 1：　風土條件 (Terroir) 指的是各村落 / 各葡萄坡的鈣含量、日照方向與風向等天然環境條件。

註 2：歐盟規定關於巴洛羅紅酒的十一個法定村落分別為：Barolo (BA), Castiglione Falletto (CF), Cherasco (CR), Diano d'Alba (DA), Grinzane Cavour (GC), La Morra (LM), Monforte d'Alba (MO), Novello (NO), Roddi (RO), Serralunga d'Alba (SE), 和 Verduno (VE). 每一個村莊中各有其著名之葡萄坡，義大利人慣稱為 "CRU"。

Barolo and its crus

One thing I love most about Barolo wine is that there's a dynamic, systematic and quite scientific match of the taste and of its terroir each crus if you are able to hold the thought of "the official map and 11 villages (P.16, No.2)" while you experience each glass of different Barolo. For sure one person will have to drink enough different Barolo in order to match and compare such connection of WINE to TERROIR, through which I highly recommend websites below:

1. www.BaroloLibrary.com   2. www.EnotecadelBarolo.it   3. www.enogea.it

One of the most interesting and challenging method to understand Barolo is through blind taste. The connection between each TERROIR and each WINE can be distinguished by its taste and different soil composition. In the case of Barolo, with different level of calcium we can roughly draw a 45° line: the North-West being "Female" while the South-East being "Male". Of course the exposition and vinification of each winery also make differences, yet it is common to find the former being elegent and with tender acidity just like female character while the latter presents more wood and leather as male. Often we hear words such as "Cherry", "Black Current", "Wood" and "Leather" when people describe Barolo.

註 3： 巴洛羅紅酒的葡萄品種名。

註 4： 櫻桃 (Cherry)、黑醋栗 (Black Current)、橡木 (Wood) 與皮革 (Leather) 等皆為用來描述、溝通葡萄酒口感與表現的專業標準用語。

北義盲飲傳統：巴洛羅葡萄酒新上市

　　根據歐盟的法規，巴洛羅紅酒 (Barolo) 和巴爾芭萊斯克紅酒 (Barbaresco) 必須於木桶陳釀至少 38 個月和 26 個月以上、葡萄品種規定為娜比歐羅葡萄種 (Nebbiolo grapes) 且產地必須於其各自規定的法定區內 [5]，始能於酒標上標示該酒名稱並得到 DOCG 的歐盟原產地認證。「巴洛羅紅酒 (Barolo)」和「巴爾芭萊斯克紅酒 (Barbaresco)」這兩大義大利名酒皆為陳年「娜比歐羅紅酒 (Nebbiolo Wine)」，兩者葡萄品種相同卻各自擁有其獨到之處。

　　每一年五月的每周五晚上，由「巴洛羅紅酒協會」和「巴爾芭萊斯克紅酒協會」舉辦的新酒發表會 (義大利文名稱為 IL Barbaresco a Tavola) 採用盲飲方式，提供一次品嘗 16 支以上的陳年娜比歐羅紅酒的機會。相同葡萄品種相同區域的葡萄酒，只因其不同風土條件 (同 P.16 註一) 與其不同釀酒師與哲學、而使其表現迥異。Il Barbaresco a Tavola 這個北義盲飲傳統的概念，可用垂直的「時間」(每周五晚上) 和平行的「空間」(全小鎮的餐廳) 與所有酒莊的新酒上市作為解釋：每一年五月的每周五晚上，於巴洛羅和巴爾芭萊斯克法定產區的小鎮內，無論當晚預訂哪一家餐廳，皆能一次品嘗到 16 支 ~19 支葡萄酒；同一天所有餐廳提供的酒皆相同，而每一週的酒則週週不同；活動持續直至所有酒莊的新酒輪過一回方才結束。在義大利這樣的盲飲傳統，每年吸引了大量的品酒師、伺酒師、酒品販售者、通路媒體記者、餐廳主廚與經理、甚至釀酒師與莊園負責人前來參加。有興趣報名的讀者亦可登錄義大利美食教育中心網站 (www.CLUBalogue.com) 尋得更多資訊。

註 5：巴洛羅紅酒的法定產區為 11 個村落，請見 P.16 註 2；巴爾芭萊斯克紅酒的法定產區為 4 個村落，見 P.19 註 6。

Il Barbaresco a Tavola

According to the law, it is necessary to age the Nebbiolo wine for at least 38 months to be "Barolo DOCG" and 26 months, "Barbaresco DOCG". The grapes must originate from the official communes of Barolo (11 villages, see P.16) and Barbaresco (4 villages; see below) in order to acquire the certification of its own DOCG (Denominazione di Origine Controllata e Garantita). Both wine, though from the same variety of grapes and both aged with certain period of time, perform differently with its own unique characters transformed from its own terroir and from the art of enologists.

The event "Il Barbaresco a Tavola" is, in my experience, one of the best ways to explore the world of Nebbiolo wines in Piemonte. The concept of this tradition is to offer opportunity for public to taste as many as possible different new Barbaresco wine in Barbaresco village every consecutive Friday of May. In this period, all restaurants serve its best cuisine no matter in tradition or in new fashion. Every Friday all restaurants serve the same 16 to 19 Barbaresco wine while there are different ones for each week; it lasts until all attending wineries are presented. For me, the idea is easily understood when located TIME (every Friday evening of May) in vertical line and SPACE (all restaurants in the town of Barbaresco but also in Treiso, Neive, Alba and more) in horizontal line, crossing with all the new Barbaresco wine from all wineries. It is truly an event worthy of participation.

註 6 ： *The 4 offical villiages are Barbaresco, Treiso, Neive, and part of the territory of the Alba municipality.*

1st Judge Tasting

— Lucca, Italy —

第一場 / 巴洛羅評審會

主評人 **Leading judge**：Fausto Borella
主　廚 **Chef**：Giuliano Pacini

評審地點 LOCATION

聖安東尼百年老酒館 Buca di Sant'Antonio
義大利托斯卡尼省盧卡城 Lucca, Toscana, Italy

餐廳的歷史

創始於西元 1782 年，擁有 200 多年歷史的〈聖安東尼百年老酒館〉，一直以來皆由帕企尼家族 (Pacini) 經營，目前擁有者為 Giuliano Pacini 和 Franco Barbieri。此餐廳保留了真正義大利托斯卡尼傳統菜餚而頗具盛名，吸引各界名人，如瑞典古斯塔羅國王以及英國瑪格麗特公主 (斯諾登伯爵夫人) 皆曾多次來訪。

Dated from 1782, Buca di Sant'Antonio belongs to Pacini family and has kept its tradition ever since. Its reputation and real Tuscany cuisine has appealed to famous people such as the King Gustavo of Sweden and the Princess Margaret of England. The owners now are Giuliano Pacini and Franco Barbieri.

主廚介紹

朱利安諾帕企尼 (GIULIANO PACINI)

身為托斯卡尼盧卡古城中最著名的傳統菜餚主廚之一，朱利安諾帕企尼於西元 2014 年榮獲義大利廚藝學院之 Nuvoletti 獎，作為其對於保留地方傳統菜餚努力之最大肯定。

One of the leading interpreters of traditional cuisine of Lucca; Knight of the Republic; in 2014 he received the award Nuvoletti from the Italian Cuisine Academy, reserved for the restaurant or organization that has contributed significantly to the conservation, knowledge and appreciation of traditional good food of their territory.

筱雯推薦的〈不可錯過〉What you should not miss!

網址 website: www.bucadisantantonio.com
地址 address: Via della Cervia, 3, 55100 Lucca, Italy
電話 tel: +39 0583 55881
推薦菜餚 Xiaowen's Recommended Dish:
托斯卡尼燉煮牛肚 (Trippa di vitello alla lucchese)
托斯卡尼家常雞蛋義大利麵佐新鮮松露刨片及熱熔奶油
(Taglioline al burro fuso e tartufo)

評審團
JUDGES

法烏多・波雷拉

義大利美食教育中心副董事、義大利最佳 60 橄欖油競賽創辦人、橄欖油評審、專業品酒師

原本承襲父業為刑事律師，然對於義大利葡萄酒和處女初榨橄欖油文化更具熱情。師承義大利知名美食評論家記者 Luigi Veronelli；生於西元 1970 年並從此全家居住於托斯卡尼盧卡古城。

Fausto Borella

Associate Director of CLUBalogue Academy, Founder of Maestrod'Olio, Oil Expert, and wine taster
Used to be a criminal lawyer following his father's footsteps, but he remained spellbound to wine and extra vergin olive oil thanks to Luigi Veronelli. Born in 1970, lives in Lucca with his family.

伊拉莉亞・蒂瑪契歐

義大利一星米其林餐廳主廚

生於佛羅倫斯的她，在蒙地卡羅師習大師如 Joël Robuchon 和 Alain Ducasse 等三星米其林主廚；回到義大利後得到了屬於自己的一星米其林榮耀於 Le Tre Lune 餐廳，最拿手項目為甜點。

Ilaria Di Marzio

1star Michelin Star Chef/Patissier, Le tre Lune Calenzano, Firenze
Born in Florence, worked in Monte Carlo (Metropole, Joël Robuchon, Le Louis XV of Alain Ducasse), then in France (Les Crayères Philippe Mille in Reims). Later returned to Italy and worked in Le Tre Lune, where she gets her first Michelin star.

范倫蒂多・卡薩內里

托斯卡尼五星級飯店行政主廚

曾經任職於英國倫敦知名高級餐廳 Locanda Locatelli 和 NOBU；回到義大利後任職於米蘭二星米其林餐廳 Cracco；出生於具多元農產文化的義大利中北部蒙特納。

Valentino Cassanelli

Executive Chef, Lux Lucis Hotel Principe Forte dei Marmi, Tuscany

His passion and his hunger for knowledge took him to London in Locanda Locatelli and NOBU. Back to Italy, he starts at "Cracco" in Milan. Valentino was born near Modena, land of great culinary tradition.

索克・貴寇

托斯卡尼五星級飯店資深伺酒師

出生於 1978 年；2006 年取得義大利伺酒師認證而 2010 年取得官方品酒師資格。熱愛美酒，尤其是法國香檳；他熱愛他的工作且生了一對雙胞胎。

Sokol Ndreko

Wine Sommelier and Maitre, Lux Lucis Hotel Principe Forte dei Marmi, Tuscany

Born in 1978, got the Sommelier qualification in 2006 and official wine taster in 2010. He loves all the good wine, especially Champange. He loves my job ; father of twin sons.

柯拉多・帕奇尼

義大利伺酒師、百年餐廳傳人

說他一輩子都在家族傳承的餐廳裡工作、實在不為過；大學畢業於音樂與戲劇系，他就是家族餐廳經營管理的「全職導演」。同時也是專業伺酒師的他，全心支持托斯卡尼菜餚與當地葡萄酒的搭配。

Corrado Pacini

Wine Sommelier and owner son, Buca di S.Antonio
Practically been working in the family restaurant since forever. After graduating in Music, Theatre and Film, he is directing full-time the room of the Buca. As professional wine sommelier, he is a strong supporter of local cuisine combined with great wines from Tuscany.

薩姆爾・柯聖提諾

托斯卡尼餐廳老闆兼主廚

生於西元 1971 年；於 1995 年與妻子 Silvia Pacini（同時也是專業伺酒師）及岳父 Giuliano Pacini（餐廳主廚）共同創立了屬於自己的餐廳至今。

Samuele Cosentiono

Chef and owner, Restaurant Gli Orti di Via Elisa, Lucca, Tuscany
Born 1971 and in 1995 acquires shares of the restaurant Gli Orti di Via Elisa, together with his father-in-law, Chef Giuliano Pacini and his wife Silvia Pacini (Wine sommelier).

安傑羅‧拓屈揚尼
托斯卡尼老字號餐廳主廚

喜歡戲稱自己為小廚而非主廚,對於美食烹飪卻是「拗」到深處無怨尤;他專研義大利傳統食譜,將不起眼的小空間轉變成為饕客滿堂的頂級餐廳;擅長於托斯卡尼傳統菜餚。

Angelo Torcigliani
Chef, Restaurant Il Merlo Camaiore, Tuscany

A chef who likes to call himself a cook, a glutton unrepentant. Born from old gastronomy, he drew a small place to gourmet restaurant. He is a great interpreter of gourmet cuisine of traditions.

利多‧范努奇
義大利酒愛好者、義大利知名美食攝影師

長久以來喜愛美食美酒,因此成為美食攝影師;曾經為釀酒師因而十分擅長於品酒;發表攝影作品於「盧卡美酒珍寶(暫譯)」與「麵包世界(暫譯)」

Lido Vannucchi
Wine lover and Food Photographer

Gourmet photographer, good food and good wine lover all along, for years he has been specializing in food photography. Expert at wine and food field with a past as a winemaker. He published photos in the volumes "Lucca Wine Treasures" and "Di Pane in Pane".

16支不同巴洛羅紅酒
16 BAROLO SELECTED

16 支不同的巴洛羅紅酒　16 Barolo selected

◆ 酒莊與編號 Winery and No.　▲ 葡萄坡名 CRU　❀ 評審會盲飲結果 Judge choices
● 酒名與年份 Wine/Vintage　■ 酒精濃度 Alcohol degree

01 p.132 ❀TOP3

◆ Castello di verduno (03_1)
● Barolo MASSARA Docg 2009
▲ Massara, Verduno
■ 14.5%vol.

05 p.149

◆ Manzone Giovanni (10_3)
● Barolo BRICAT Docg 2010
▲ Gramolere, Monforte d'Alba
■ 14.5%vol.

02 p.134

◆ Cascin Adelaide (04_1)
● Barolo FOSSATI Docg 2010
▲ Fossati, La Morra
■ 14%vol.

06 p.150 ❀TOP5

◆ Borgogno Francesco (11)
● Barolo BRUNATE Docg 2010
▲ Brunate, La Morra
■ 14.6%vol.

03 p.139 ❀TOP2

◆ Bruna Grimaldi (05_1)
● Barolo BADARINA Docg 2010
▲ Badarina, Serralunga d'Alba
■ 14.5%vol.

07 p.153

◆ Claude Boggione (14)
● Barolo BRUNATE Docg 2010
▲ Brunate, Barolo
■ 14.5%vol.

04 p.144

◆ Giribaldi (07)
● Barolo GIRIBALDI Docg 2010
▲ Ravera, Novello
■ 14%vol.

08 p.154

◆ Ettore Fontana-Livia Fontana (15)
● Barolo VILLERO Docg 2009
▲ Villero, Castiglione Falletto
■ 13.5%vol.

◆ Alario Claudio (16)
● Barolo SORANO Docg 2010
▲ Sorano, Serralunga d'Alba
■ 14.5%vol.

◆ Tenuta Rocca (30)
● Barolo BUSSIA Docg 2010
▲ Bussia, Monforte d'Alba
■ 14%vol.

◆ Ca'Rome' (18_1)
● Barolo RAPET Docg 2010
▲ Cerretta, Serralunga d'Alba
■ 14%vol.

◆ Cadia (35)
● Barolo MONVIGILIER Docg 2010
▲ Monvigliero, Verduno
■ 14.67%vol.

 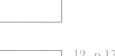 TOP1

◆ Bric Cenciurio (20)
● Barolo COSTE DI ROSE Docg 2011
▲ Coste Di Rose, Barolo
■ 14%vol.

TOP6

◆ Paolo Manzone (36_1)
● Barolo MERIAME Docg 2011
▲ Meriame, Serralunga d'Alba
■ 14%vol.

TOP4

◆ Dosio Vigneti (27_1)
● Barolo FOSSATI Docg 2009
▲ Fossati, Barolo
■ 14%vol.

◆ Sanbiagio (39_1)
● Barolo SORANO Docg 2010
▲ Sorano, Serralunga d'Alba
■ 14%vol.

給讀者的話
JUDGES COMMEND

主評人 / 法烏多 · 波雷拉

生於托斯卡尼且為葡萄酒之專業人士，我們
習慣且偏愛托斯卡尼葡萄酒及品種（聖爵維斯
葡萄種），因此對我們來說，一天之中品嘗 16
種來自北義的巴洛羅紅酒，絕對是一大最難
得而不易的經驗。這些酒年輕、具適當單寧
且可口，新鮮而引人入勝的酸度正為娜比歐
羅葡萄種最重要的特徵。廚師、伺酒師、品
酒師、以及美食領域的專業人士們共聚一堂，
仔細品嘗每一瓶酒、並分析其口感。不同於
二十世紀的巴洛羅，這些二十一世紀的新巴
洛羅更顯層次。最後，所有來自托斯卡尼的
評審全體同意以下：這十六瓶酒皆來自於絕佳

的風土環境 (terroir)、經由資深釀酒師具長年經驗的雙手、將生長於優質土質的
葡萄轉化成為、那擁有完美表現的「義大利巴洛羅紅酒」。

Leading Judge / Fausto Borella

For wine professional sommeliers born and raised with traditional Tuscan wines,
tasting 16 Barolo wines in one day means THE unique experience. Young, tannic,
mellow, with a fresh acidity, some still unexpressed while others ready to be tasted.
Each Barolo was being tasted carefully and stripped of all taste and smell clothing.
Chef, sommelier and maître, the present judges, have offered their tasters experience
and defined these wines as they occurred in tasting: forthright, determined, with a
different twist from older Barolo.

For once, all present tasters agreed with the wine of the day, proving that when
the grapes are grown in the right ground and transformed by the expert hands of
winemakers who respect the terroir, the result is inevitable: a Barolo of excellence.

2nd Judge Tasting

— Piemonte, Italy —

第二場 / 巴洛羅評審會

主評人 Leading judge：Stefano Dal Ry
主　廚 Chef：Marco Sacco

評審地點 LOCATION

二星米其林餐廳 Piccolo Lago/ 2-star Michelin Restaurant
義大利皮爾蒙特省維爾巴尼亞 Verbania, Piemonte, Italy

餐廳的歷史

"Piccolo Lago" 義大利文直譯中文為「小湖」；顧名思義，餐廳位在北義湖區的小湖旁。在此處用餐若餐桌接近落地窗，窗外盡是湖光山水，陽光灑落餐桌更如行船於湖上。餐廳旁設有停機棚、可供搭乘直升機的客人降落及起飛。

"Piccolo lago" means "little lake" and the name itself has best described its location and atmosphere which is breathe-taking and irreplaceable. If your table is close to the window, you may bathe in the beauty of the lake; while the sunshine pays upon, you might feel as well as in a boat over the lake.

主廚介紹

馬可・薩柯是北義皮爾蒙特省重要的主廚之一。他的料理從皮爾蒙特的食材傳統出發，發展出獨特且新創的料理哲學。他善用位於北義湖區之地緣之利，將湖魚、山牛與鮮蔬之美味搭配各式草本香料。其美味而有趣的菜餚很快地得到米其林二星的肯定。

Marco Sacco is one of the most important chefs in Piemonte. His cuisine is based on Piemonte tradition to develope his own philosophy of cuisine. Chef Marco matches local ingredients such as lake fish, Piemonte beef, fresh vegetables and herbs with oriental spices which in combination, creative and interesting dishes.

筱雯推薦的〈不可錯過〉What you should not miss!

網址 website: www.piccololago.it
地址 address: Via Filippo Turati 87, 28924, Verbania VB, Italy
電話 tel: +39 0323 586792
推薦菜餚 Xiaowen's Recommended Dish:
<與馬可共廚>套餐：23 道主廚親手製作的菜餚
(Twenty - three tastings "In the kitchen with Marco")

評審團
JUDGES

馬可・薩柯
義大利二星米其林餐廳主廚

身為二星米其林主廚的他、特別注重「食材與土地」的關係，尤其是從旅行當中得到靈感；然無論如何，皮爾蒙特省的家鄉味，對他來說依舊是世界上最棒的美食美酒，尤其是巴洛羅紅酒。

Marco Sacco
2star Michelin chef, Piccolo Lago, Piemonte

Two Michelin stars Chef with a particular attention to territories and their ingredients. He likes to take inspiration from his journeys through the world but he loves the best the taste of his Piemonte... Barolo included !

法蘭可・馬拉斯科
北義知名高端創意料理餐廳主廚兼老闆

極具新意及靈感的義大利創意主廚；他同時對非洲料理十分擅長 (因為老婆的娘家在非洲)；他喜愛海鮮料理，無論魚種來自海洋或是家鄉馬焦雷湖；魚之於水、如同他之於美酒，如果是巴洛羅紅酒更好！

Franco Marasco
Chef owner, Il Clandestino in Stresa, high level restaurant with creative cuisine

Really creative and clever in his chef job, he knows very well African kitchen tradition because his wife is from Africa. He likes cook sea fish but also fish from Maggiore lake. Water for fish and wine for him... if Barolo better !

佛朗西斯・米諾・米諾拉

北義四星級飯店行政主廚

來自於義大利南部的他，結合北義當地食材與南部傳統烹調方法，在美麗的義大利綺法飯店 (Ghiffa)，每每創造出令人難忘的美味。巴洛羅紅酒在他心中的葡萄酒名單、名列前茅。

Francesco Miro Mirolla

Executive Chef, 4-star hotel Ghiffa

Being chef in a beautiful hotel in Ghiffa on lago Maggiore, he has south of Italy traditions in his idea of kitchen and this gives a special sensation to his plates that remains in your mind. Barolo is on top of his personal wine list.

盧卡・莫利納

托斯卡尼五星級飯店資深伺酒師

他本來是一位在羅馬的執業律師，但現居住於北義馬焦雷湖且從事專業伺酒師的工作；當時的他、來到北義之後便愛上了葡萄酒的世界。身為義大利伺酒師協會楓巴尼亞省伺酒師代表的他，時常舉辦品酒課程，當然，巴洛羅紅酒活在他的心尖上。

Luca Molina

Wine Sommelier, Verbania province Delegate

He lives on Maggiore Lake but he comes from Rome. He was a lawyer, but when he arrived in Piemonte he discovered a great love for wine! As Sommelier and AIS Delegate, he organizes tasting wine courses. Barolo in his heart.

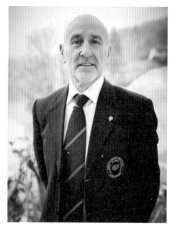

保羅・費拉羅

義大利伺酒師協會北部區域理事長
暨楓巴尼亞省伺酒師代表

身為義大利酒師協會北部區域理事長暨地方代表，葡萄酒之於他的人生，想當然爾，他的品酒經驗與技巧不需多言，然一瓶新的巴洛羅紅酒卻總是能帶給他美妙的驚喜。

Paolo Ferraro

Wine Sommelier, Verbania province Delegate
Wine is in his life since ever, he is the President of AIS in the north Piedmont area and we surely can say he has a very long experience in wine tasting. Always searching quality surprises opening a new bottle... first of all Barolo.

里卡爾多・米蘭

北義餐飲學院教授及知名品酒記者

他是北義餐飲學院教授，同時也是知名品酒記者，對葡萄酒可用「狂愛」來形容。使用簡而易懂的文字、風趣而調侃的態度，是他欣賞葡萄酒、獨具風格的途徑。如果你問：「來杯巴洛羅紅酒嗎？」他會帶著他一貫的笑容告訴你：「這真是個好主意」。

Riccardo Milan

Wine journalist and Stresa Scuola Alberghiera teacher
Journalist, teacher in Stresa Hotelier School, blogger and wine enthusiast, he has a very simple and interesting way to talk about wine and about what people say around wine! Clever and funny in his way to appreciate wine... Barolo? Good idea guys!

辛西亞·費羅

義大利伺酒師協會楓巴尼亞省伺酒師、義大利
Tavola 競賽調酒冠軍、北義知名酒吧老闆娘

能幹的酒吧老闆娘、同時是個調酒冠軍，她對於全
義大利的酒如數家珍。「葡萄酒」對她來說，是工
作也是她心中重要的一個角落。如果缺乏巴洛羅紅
酒，她的酒吧可是不能開門營業的呢！

Cinzia Ferro

Wine Sommelier, Winner of national contest on
Tavola magazine, owner of Estremadura bar
Great barlady, passionate about her job and well-
known all over Italy, she also received a national price
from a specialized magazine. Wine is part of her job
and it takes part of her heart. Her bar cannot open
without Barolo.

斯特凡諾·丹爾瑞

義大利酒愛好者、義大利知名美食公關公司
創辦人兼股東

餐飲公關、美食、廚師，是他的工作內容也是興趣；
他說一天之中如果沒有一杯美酒，堪稱完美中的遺
憾，若這杯美酒是來自拉莫拉產區、他的酒莊好友
群中所生產的巴洛羅紅酒，則人生無所求！

Stefano Dal Ry

Wine lover and owner of gourmet shop and ICS
communication company
Communication, restaurant and food in general are
main part of his job and interests. He says it's not a
really good day without at least a glass of good wine...
if Barolo, made in La Morra by some friends of him,
better!

16 支不同巴洛羅紅酒
16 BAROLO SELECTED

◆酒莊與編號 Winery and No.　　▲葡萄坡名 CRU　　　　 評審會盲飲結果 Judge choices

●酒名與年份 Wine/Vintage　　■酒精濃度 Alcohol degree

01　p.130
◆ Aurelio Settiomo (01)
● Barolo ROCCHE dell' Annunziata Docg 2009
▲ Frazione Annunziata, La Morra
■ 14%vol.

05　p.147
◆ Manzone Giovanni (10_1)
● Barolo CASTELLETTO Docg 2010
▲ Castelletto, Monforte d'Alba
■ 14.5%vol.

02　p.131
◆ Bersano (02)
● Barolo BADARINA Docg 2008
▲ Badarina, Serralunga d'Alba
■ 14%vol.

06　p.159　 TOP5
◆ Massolino (19_2)
● Barolo PARUSSI Docg 2010
▲ Parussi, Castiglione Falletto
■ 14.5%vol.

03　p.135　 TOP2
◆ Cascina Adelaide (04_2)
● Barolo PERNANNO Docg 2010
▲ Pernanno, Castiglione Falletto
■ 14.5%vol.

07　p.172
◆ Dosio Vigneti (27_2)
● Barolo Docg 2010
▲ Mix, La Morra
■ 14%vol.

04　p.140　 TOP1
◆ Bruna Grimaldi (05_2)
● Barolo CAMILLA Docg 2010
▲ Raviole, Grinzane Cavour
■ 14%vol.

08　p.188
◆ Marchesi Di Barolo (38_3)
● Barolo COSTE DI ROSE Docg 2011
▲ Coste Di Rose, Barolo
■ 14.5%vol.

09 p.190 TOP4

◆ Sanbiagio (39_2)
● Barolo SAN BIAGIO Docg 2010
▲ San Biagio, La Morra
■ 14%vol.

13 p.205 TOP6

◆ Palladino (49_1)
● Barolo ORNATO Docg 2009
▲ Ornato, Serralunga d'Alba
■ 14.5%vol.

10 p.192

◆ Fratelli Giacosa (40)
● Barolo BUSSIA Docg 2010
▲ Bussia, Monforte d'Alba
■ 13.5%vol.

14 p.208

◆ Conterno Fantino (55_1)
● Barolo Sorì Ginestra Docg 2011
▲ Ginestra, Monforte d'Alba
■ 15%vol.

11 p.193

◆ Scarzello Giorgio e Figli (41)
● Barolo SARMASSA VIGNA MERENDA Docg 2008
▲ Sarmassa Vigna Merenda, Barolo
■ 14.5%vol.

15 p.220

◆ Attilio Ghisolfi (74_1)
● Barolo BUSSIA Docg 2010
▲ Bussia, Monforte d'Alba
■ 14.5%vol.

12 p.203 TOP3

◆ Ettore Germano (48_1)
● Barolo CERRETTA Docg 2010
▲ Cerretta, Serralunga d'Alba
■ 14.5%vol.

16 p.225

◆ Abbona (85_1)
● Barolo Pressenda 2010
▲ Castelletto, Monforte d'Alba
■ 14%vol.

給 讀 者 的 話
JUDGES COMMEND

主評人 / 斯特凡諾 · 丹爾瑞

對所有的評審來說，這一次盲飲 16 瓶巴洛羅
紅酒的經驗，可說是非常棒也非常困難！儘管
評審們平日時常品評各酒莊的葡萄酒，但當
每一次盲飲的結果，顯示知名酒莊不一定表
現最好的時候 (雖然時常發生這樣的情況)，
總還是令我們驚訝不已！品飲葡萄酒需要使用
到的技巧不只是感官經驗，更是身為單一個
體對於「生命瞬間」的完整熱情表現，當每
一個評審參與品評的當下，理性與科學技巧
只是一部分，更重要的是當下對於葡萄酒的
喜愛與享受，尤其是當盲飲不知杯中之酒為
何酒莊時，最後得到結果的驚奇感，能令人
對葡萄酒的態度 (甚至是人生態度) 有所改觀。

特別謝謝筱雯邀請我們參加、她在二星米其林餐廳 Piccolo Lago 舉辦的巴洛羅紅
酒評審會！

Leading Judge / Stefano Dal Ry

Really nice experience and really a hard job to taste so many different Barolo!
Always surprising to see that, as many times happen, the best winery names are not
on the top of the classification. We can say that drinking wine is not only a sensory
experience but a complete moment dedicated to a passion and everybody join it
acting as a fan or a supporter... rationality and technical elements are only a part of
the normal judgment of wine! Without seeing the label on the bottle, all this changes
and results are always surprising.
Thanks Xiaowen for this nice Barolo day at Piccolo Lago restaurant!

3rd Judge Tasting

— Piemonte, Italy —

第三場 / 巴洛羅評審會

主評人 Leading judge：Marcella Bianco
主　廚 Chef：Alessandra Buglioni di Monale

評審地點 LOCATION

義大利維杜農百年城堡 Castello di Verduno
義大利皮爾蒙特省維杜農 Verduno, Piemonte, Italy

餐廳的歷史

維杜農百年城堡是巴洛羅紅酒歷史文化中，最重要的地點之一。自 1838 年起，義大利第一任國王維托里奧・埃馬努埃萊二世任命其釀酒師卡羅・史達葉諾於此地專為宮廷貴族釀酒，便開啟了維杜農城堡輝煌的巴洛羅紅酒文化歷史。

Castello di Verduno is one of the most important castles in the history of Barolo culture. Starting from 1838 when Vittorio Emanuele II was the first King of Italy, the enologist General Carlo Staglieno started here his work for the Royalty and thus, the beginning of Barolo culture in Castello di Verduno.

家族介紹

維杜農百年城堡的產業目前是由碧揚卡和本羅多兩大家族共同傳承與管理，包含兩大酒莊、兩家餐廳、與兩家旅館。法蘭可・碧揚卡與嘉碧雅菈・本羅多兩人浪漫的愛情故事，使得維杜農城堡成為同時擁有義大利兩大名酒的指標性酒莊。

Bianco Family fully operates Castello di Verduno: there are the historical winery, 2 restaurants, and 2 hotels. Franco Bianco, married to Gabriella Burlotto, works in the winery while the sister, in the tourism of the castle. Alessandra Buglioni di Monale, the daughter-in-law, and Giovanna Bianco, the youngest daughter of Franco Bianco, are now the Executive Chefs of each restaurant.

筱雯推薦的〈不可錯過〉What you should not miss!

網址 website: www.castellodiverduno.com
地址 address: Via Umberto I, 9, 12060 Verduno, Italy
電話 tel: +39 0172 470125
推薦菜餚 Xiaowen's Recommended Dish:
皮爾蒙特傳統燉牛舌佐紅醬 (Lingua di vitello in salsa rossa)、手作北義肉醬雞蛋麵 (Tagliatelle al ragù)、義大利奶酪 (Panna Cotta)

評審團
JUDGES

瑪雀拉・碧揚卡

北義帕拉維嘉葡萄酒協會理事長、維杜農城堡主人

我熱愛美酒、我的工作、我的土地；如此的熱情來自於我的家族傳承，我與我的家族成員一同工作著；畢業於食品科學釀酒科系，我致力將美酒美食的正面能量，帶給所有來到維杜農城堡的人們。

Marcella Bianco

President of Pelaverga Association and Owner of Castello di Verduno

I am a girl who loves wine, loves her job and loves her land. The passion was passed to me by my family that I work with. Degree in enology in University of Gastronomic Sciences, now I try to do my best in my wine and to sensitize my customers to "good".

沃特・斐雷投

義大利一星米其林餐廳主廚

我生於廚師餐飲世家；自 1985 年我於此立根後數年，我便得到了米其林一星的肯定。如果我不是廚師，我想我必定沉浸於經濟與財經的世界；我喜歡跟朋友享受美食美酒；我喜歡旅行、學習他國文化。

Walter Ferretto

1star Michelin Chef, Il Cascinalenuovo, Piemonte

I was born and grow up in a family of restaurants. In 1985 landing Il Cascinalenuovo and after few years I get the Star on Michelin Guide. If I had not become chef, I loved the economic & financial world. I love spend time with my friends eating and drinking fine wines and I like travel to learn traditions of other countries.

馬立歐 · 亞吉歐翁

維杜農城堡酒莊釀酒師

生於杜林市中心然心近大自然、畢業於杜林釀酒師學院；至今於義大利葡萄酒領域工作長達 15 年，其全方位的技能包含葡萄樹生長與土壤、釀酒、市場研究、酒莊經營管理、宣傳與公關。

Mario Adrion

Enologist, Castello di Verduno, Piemonte

Born in the city of Turin but he always felt more comfortable into the nature, degree in Viticulture and Enology at the Faculty of Agronomy of Turin. He is working in the wine business for more than 15 years. Skills at 360°: vinegrowing, winemaking, marketing, managing, advising and public relation.

亞拉珊卓 · 布依翁尼

維杜農城堡餐廳主廚

我的祖父 Giovanni Battista Burlotto 於 1953 年開創了維杜農城堡餐廳，之後傳給我的母親、再傳給我。我在我的菜餚中致力於創造其好吃的口感，帶給人們感官上的享受、以及情感與料理哲學想法上、建立與之其原食材的溝通橋樑。

Alessandra Buglioni di Monale

Executive Chef, Real Castello di Verduno, Piemonte

My mom was the chef of the restaurant at Real Castello di Verduno that my grandfather Giovanni Battista Burlotto opened in the 1953. I am dedicated to food that are taste sensations, emotions that they procure, communicability and the physicality of the ingredients.

蘿拉・梅納蒂
義大利伺酒師協會皮埃蒙特省伺酒師

40 歲、自 2003 年起因熱愛葡萄酒而於阿爾巴擔任專業伺酒師、並時常協助當地的葡萄酒展覽。她嫁給了個性爆笑且同為伺酒師的先生；目前有一個 2 歲的小嬰兒且她認為她的小孩以後也會很喜愛葡萄酒。

Laura Meinardi
Wine Sommelier

40 years old, Sommelier for passion in Alba from 2003, works as shopkeeper and Sommelier in Alba restaurants and assists exibitions. Married with a funny sommelier, mother of a 2 year-old baby that will enjoy wines a lot in his future.

法德利克・法瑞羅
義大利伺酒師協會皮埃蒙特省伺酒師

41 歲、自 2000 年起因熱愛葡萄酒而於阿爾巴取得專業伺酒師認證；他的正職為義大利汽車公司工程師；目前與蘿拉 梅納蒂擁有一個小嬰兒、有著美好的婚姻生活。他認為「葡萄酒將永遠存在於我們三人歡愉的時刻」。

Federico Ferrero
Wine Sommelier

41 years old, Sommelier for passion in Alba from 2000, mechanical engineer in automotive company for work and happily married with Laura and father o the same lucky baby. He says, "wines have been anc will be always with us during our pleasant moments."

吉歐・馬特

義大利知名部落客

天生熱愛關於義大利的一切、對於新潮且不凡的事物充滿追求；是個美食旅行者和攝影紀錄者、是專業伺酒師也是品酒師；致力於發現義大利之美且用攝影來記錄「他」的美妙經驗。

Gio Marta

Blogger, Italy

Born and raised with love for all italian. A hunter for cool and unusual. A traveller and photographer, sommelier and connoisseur. Strive to discover the wonders of italy and recount the experiences in photography.

瑪蓮娜・德加

義大利伺酒師協會皮埃蒙特省伺酒師

35 歲、已婚、波蘭人但自 2001 年起便居住於義大利。她熱愛旅行與品酒；自 22 歲第一次品酒開始，她便立志成為專業伺酒師。她時常參與葡萄酒會並誓志永不停止學習葡萄酒的這條路。

Marlena Deja

Wine Sommelier

35 years old, married. From Poland but lives in Italy since 2001. She love traveling and wine tasting. Age 22 she started to taste her first wines and becomes sommelier. She participates in several events and never stops learning about wine and discovers new labels and flavours.

16 支不同巴洛羅紅酒
16 BAROLO SELECTED

16 支不同的巴洛羅紅酒　16 Barolo selected

◆ 酒莊與編號 Winery and No.　▲ 葡萄坡名 CRU　　　 評審會盲飲結果 Judge choices
● 酒名與年份 Wine/Vintage　　■ 酒精濃度 Alcohol degree

__01__ p.138

◆ Cascina Adelaide (04_5)
● Barolo BAUDANA Docg 2011
▲ Baudana, Serralunga d'Alba
■ 14.5%vol.

__05__ p.157
TOP2

◆ Ca 'Rome' (18_2)
● Barolo CERRETTA Docg 2010
▲ Cerretta, Serralunga d'Alba
■ 13.5%vol.

__02__ p.142

◆ Sebaste Mauro (06_1)
● Barolo Trèsüri Docg 2010
▲ Mix, Grinzane Cavour
■ 14%vol.

__06__ p.164 TOP4

◆ Sordo Giovanni (22_3)
● Barolo PERNO Docg 2011
▲ Perno, Monforte d'Alba
■ 14.5%vol.

__03__ p.145

◆ Rivetto (08)
● Barolo del Comune di Serralunga 2010
▲ Mix, Serralunga d'Alba
■ 14.5%vol.

__07__ p.179 TOP6

◆ Gemma (32)
● Barolo COLAREJ Docg 2011
▲ Collaretto, Serralunga d'Alba
■ 14%vol.

__04__ p.151

◆ Sylla Sebaste (12)
● Barolo BUSSIA Docg 2010
▲ Bussia, Barolo
■ 14%vol.

__08__ p.180
TOP3

◆ Vietto (33)
● Barolo PANEROLE Docg 2011
▲ Panerole, Novello
■ 14.5%vol.

_09_ p.186

- ◆ Marchesi Di Barolo (38_1)
- ● Barolo CANNUBI Docg 2011
- ▲ Cannubi, Barolo
- ■ 14.5%vol.

_13_ p.207

- ◆ Oddero Poderi e Cantine (50)
- ● Barolo Docg 2010
- ▲ Various Township
- ■ 14.5%vol.

_10_ p.194

- ◆ Josetta Saffirio (43)
- ● Barolo Docg 2011
- ▲ Mix, Monforte d'Alba
- ■ 14.5%vol.

_14_ p.209

- ◆ Conterno Fantino (55_2)
- ● Barolo VIGNA DEL GRIS Docg 2011
- ▲ Ginestra Vigna Del Gris, Monforte d'Alba
- ■ 14.5%vol.

_11_ p.195

- ◆ Ciabot Berton (44_1)
- ● Barolo ROGGERI Docg 2009
- ▲ Roggeri, La Morra
- ■ 14.5%vol.

_15_ p.221

- ◆ Attilio Ghisolfi (74_2)
- ● Barolo BRICCO VISETTE Docg 2010
- ▲ Bussia Bricco Visette, Monforte d'Alba
- ■ 14.5%vol.

_12_ p.204 TOP1

- ◆ Ettore Germano (48_2)
- ● Barolo PRAPÒ Docg 2010
- ▲ Prapò, Serralunga d'Alba
- ■ 14.5%vol.

_16_ p.226 TOP5

- ◆ Abbona (85_2)
- ● Barolo TERLO RAVERA Docg 2010
- ▲ Ravera, Novello
- ■ 14%vol.

給讀者的話
JUDGES COMMEND

主評人 / 瑪雀拉 · 碧揚卡

這是一個瘋狂的台灣女孩和巴洛羅紅酒的故事。

我認識筱雯有好幾年的時間了，從數年前我們兩人一起品飲第一杯巴洛羅紅酒、一直到現在和一群專業人士品飲十六杯巴洛羅紅酒，她做每件事情的目的，總是抱著希望、能對義大利外的世界解釋並說明「食品分級」與「品質」的重要性；在巴洛羅紅酒中，特別是葡萄酒的風土條件、製造過程、口感層次與分級。

現在的她、在世界的另一端——台灣，開始了這個瘋狂的計畫，然我相信很快地、她與她正確的食品分級與品質觀念，將會「入侵」全世界。

Leading Judge / Marcella Bianco

Xiaowen Huang, a crazy Taiwanese girl and her crazy Barolo project.

I have known her for several years from the first glass of Barolo together back then to 16 glasses of Barolo with a group of professionals now. Her goal is to explain to the world the importance of food quality, with a special focus on wine while in Barolo: he land, the grapes, the production, the idea, the taste and the classification. Now she has started this 《Barolo Library》 project only in Taiwan, but I believe she will "invade the planet" with the right food and wine culture.

4th Judge Tasting

— Piemonte, Italy —

第四場 / 巴洛羅評審會

主評人 **Leading judge**：Federico Scazello
主　廚 **Chef**：Luca Zecchin

評審地點 LOCATION

二星米其林餐廳 Guido da Costigliole/ 2-star Michelin Restaurant
義大利皮爾蒙特省聖泰法諾貝爾博 Santo Stefano Belbo, Piemonte, Italy

餐廳的歷史

餐廳開業於西元 1961 年，由古義多 · 阿爾卡提以及其妻共同經營。2002 年遷移至現址，由莫妮卡、安德亞及盧卡共同經營。其於 17 世紀為修道院、19 世紀更曾為魯意居公爵私人居住城宅，堪稱當時最負盛名的世外桃源；現結盟 Relais & Chateaux 義大利最頂級國際旅館集團。

Opened in 1961 from Guido Alciati and his wife Lidia. In 2002, the Guido da Costigliole restaurant, now led by Monica, Andrea and Luca, moved to Relais San Maurizio, a 17th century monastery founded by Franciscan monks in 1619 until 1862 it became private residence of Count Luigi Incisa Beccaria. Now it is part of the prestigious Relais & Chateaux hotel chain.

主廚介紹

盧卡 · 柴金從小師廚莉地雅·阿爾卡提，是阿爾卡提家族中的成員之一，也是國際上頗負盛名的義大利二星米其林主廚。他的烹調風格在於追求國際新潮詮釋下的義大利食材原味。

Luca was trained by Lidia Alcati from an early age it was him that Lidia put her trust of Alcati Family. Under his guidance the restaurant has garnered Micheline stars and international praise. Luca's experienced cooking blends Piemontese flavors with international ideas. His cuisine while deeply rooted in Piemonte is always refreshingly contemporary.

筱雯推薦的〈不可錯過〉What you should not miss!

網址 website: www.guidosanmaurizio.com
地址 address: Localita'San Maurizio,
　　　　　　　12058 Santo Stefano Belbo CN, Italy
電話 tel: +39 0141 844455
推薦菜餚 Xiaowen's Recommended Dish:
主廚手做義大利餃 (Revioli a Mano)、所有甜點 (All dessert)、私房酒窖 (wine collection)。

評審團
JUDGES

法德里克・斯卡柴諾

義大利官方巴洛羅紅酒協會理事長

身為巴洛羅紅酒協會 (代表超過 120 家巴洛羅酒莊) 的理事長兼釀酒師；畢業於杜林與阿爾巴的釀酒大學，擁有自己的巴洛羅酒莊，日前新添一女，與其妻居住於巴洛羅。

Federico Scazello

President of Enoteca di Barolo

As the President of Enoteca di Barolo, he represents more than 120 Barolo wineries；Graduated from Oenological Wine School in Alba and University in Turin, he owns Scarzello Giorgio & Figli Winery; Father of one daughter.

盧卡・柴金

義大利二星米其林主廚

他的烹調風格追求國際新潮詮釋下的義大利食材原味，講究北義皮爾蒙特省的在地傳統口味，然其菜餚總令人有著煥然一新的現代感。

Luca Zecchin

2star Michelin Chef, Ristorante Guido da Costigliole

His cuisine is Piemontese flavors with international ideas, deeply rooted in Langhe yet always refreshingly contemporary.

馬可・卡柏

北義知名釀酒師、酒莊老闆

身為北義最佳麝香葡萄氣泡酒的釀酒師之一,他卻
深愛著娜比歐羅葡萄。他風趣且正面的人生態度、
使得身邊的朋友總是微笑著。育有 2 子。

Marco Capra

Enologist and owner of winery

He makes one of the best Moscato in Piemonte yet
Nebbiolo grape is on his favorite wine list. His funny
and positive attitude makes people around him smile.
Father of 2 children.

安德亞・阿爾卡提

專業品酒師、米其林餐廳傳人

身為阿爾卡提家族中最小的兒子,他承接父志繼續
擴充其餐廳的酒窖規模,目前其酒藏橫跨世界三千
家酒莊、共超過三萬瓶酒,然他的最愛依舊是北義
巴洛羅紅酒。

Andrea Alciati

Wine Taster, owner of Ristorante Guido da Costigliole

Youngest of Alcati family, continues his father's
cantina from 1960's, included over 30,000 bottles and
features over 3,000 wines. Even with an international
approach, his favorite bottles remain Piemontese.

莫妮卡・瑪妮妮
專業餐飲經理、米其林餐廳股東

她是二星米其林餐廳 Guido 轉化成為北義最頂級餐廳代名詞的重要關鍵人物；她熱愛旅行、時常自我挑戰、改變，然不忘其本。

Monica Magnini
Manager and owner of Ristorante Guido da Costigliole

She transformed Ristorante Guide as the benchmark for fine dining of Langhe. In Love with travelling, constantly challenging but never forgetting tradition.

賈旭華・艾森豪爾
義大利伺酒師協會皮埃蒙特省伺酒師

美國人、畢業於紐約大學；其後於紐約擔任專業伺酒師 7 年而移居到義大利。目前為二星米其林餐廳的首席伺酒師。新婚；對於自己在北義朗格樂（巴洛羅紅酒產區）的人生滿意無比。

Joshua Eisenhauer
Wine Sommelier

Born in the United States, Josh is a graduate of New York University. Worked as a sommelier 7 years in New York City and 2 years in Piemonte. Recently married and throughly enjoying life in Langhe, Italy.

芭爾巴拉・索都

義大利知名室內設計師、時裝店老闆與食品採購

結合義大利皮爾蒙特省（母）與薩丁尼亞島（父）的傳統，是諸多義大利知名服飾店、酒吧、餐廳等商用空間的首席設計建築師，同時是義大利冠軍犬的馴獸訓練師，贏得多次的歐盟競賽；育有三子，認為食品教育以及愛護地球的綠能飲食能夠改變世界的未來。

Barbara Soddu

Architech, owner of outlet store, and Food manager
Combined ancient traditions of Sardinia (Father) and Piedmonte (mother), leading architect for Italian restaurants, raises Italian champion dogs, mother of 3 children. She likes the idea that educating children that green food can change the future of the world.

法布利奇歐・亞斯可利

義大利酒愛好者、運輸公司老闆

愛好葡萄酒之外，也曾經營過自己的餐廳，從兒時有記憶起，餐桌上的葡萄酒就是由家中父親自行釀造。他與葡萄酒的日常關係是「北義男人與葡萄酒」的最傳統典範。

Fabrizio Ascheri

Wine lover and owner of transportation company
Great lover of wine and former restaurant owner, born and raised in a family that has been producing wine for their daily use. His relationship with wine is the perfect example of Piemonte tradition: MEN & WINE.

16 支不同巴洛羅紅酒
16 BAROLO SELECTED

16 支不同的巴洛羅紅酒　16 Barolo selected

◆ 酒莊與編號 Winery and No.　▲ 葡萄坡名 CRU　　🌼 評審會盲飲結果 Judge choices
● 酒名與年份 Wine/Vintage　■ 酒精濃度 Alcohol degree

__01__ p.136

◆ Cascina Adelaide (04_3)
● Barolo PREDA Docg 2010
▲ Preda, Barolo
■ 14%vol.

__05__ p.162

◆ Sordo Giovanni (22_1)
● Barolo GABUTTI Docg 2011
▲ Gabutti, Serralunga d'Alba
■ 14.5%vol.

__02__ p.141

◆ Bruna Grimaldi (05_3)
● Barolo BRICCO AMBROGIO Docg 2010
▲ Bricco Ambrogio, Roddi
■ 14.5%vol.

__06__ p.168

◆ Le Ginestre - Grinzane Cavour (24)
● Barolo Sotto Castello Docg 2008
▲ Sotto castello, Novello
■ 14.5%vol.

__03__ p.148

◆ Manzone Giovanni (10_2)
● Barolo Gramolere Docg 2010
▲ Gramolere, Monforte d'Alba
■ 14.5%vol.

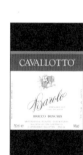

__07__ p.178

◆ Cavallotto Tenuta Bricco Boschis (31_2)
● Bricco Boschis Boschis Docg 2010
▲ Bricco Boschis, Castiglione Falletto
■ 14%vol.

__04__ p.152

◆ Silvano Bolmida (13)
● Barolo BUSSIA Docg 2010
▲ Bussia, Monforte d'Alba
■ 14.5%vol.

__08__ p.184

◆ PAOLO MANZONE (36_2)
● Barolo Serralunga Docg 2011
▲ Mix, Serralunga d'Alba
■ 13.5%vol.

09 p.187

◆ Marchesi Di Barolo
(38_2)
● Barolo SARMASSA
Docg 2010
▲ Sarmassa, Barolo
■ 14%vol.

13 p.210 TOP1

◆ Conterno Fantino(55_3)
● Barolo Mosconi Docg
2011
▲ Mosconi, Monforte
d'Alba
■ 14.5%vol.

10 p.191

◆ Sanbiagio (39_3)
● Barolo ROCCHETTEVINO
Docg 2010
▲ Rocchettevino, La
Morra
■ 14%vol.

14 p.214

◆ Giovanni Rosso (66_1)
● Barolo CERRETTA
Docg 2011
▲ Cerretta, Serralunga
d'Alba
■ 14%vol.

11 p.200

◆ Luigi Baudana-G.D.varja
(46_3)
● Barolo BAUDANA
Docg 2011
▲ Baudana, Serralunga
d'Alba
■ 14.5%vol.

15 p.217 TOP2

◆ Luigi Pira (70_1)
● Barolo VIGNARINDA
Docg 2010
▲ Vignarinda, Serralunga
d'Alba
■ 14.5%vol.

12 p.206 TOP4

◆ Palladino (49_2)
● Barolo PARAFADA
Docg 2009
▲ Parafada, Serralunga
d'Alba
■ 14%vol.

16 p.224 TOP5

◆ Giacomo Fenocchio
(75_3)
● Barolo CASTELLERO
Docg 2011
▲ Castellero, Barolo
■ 15%vol.

給 讀 者 的 話
JUDGES COMMEND

主評人 / 法德里克・斯卡柴諾

身為義大利巴洛羅紅酒協會理事長,我認為
〈Barolo Library 餐桌上的義大利酒王〉是亞
州從所未有的最重要葡萄酒活動之一,其規
模與構想來自於筱雯堅毅的熱情與專業的規
劃執行。我與義大利重要的米其林主廚、專
業伺酒師們,共同大力支持並同意參與在義
大利重要的地標進行評審活動。雖然我們在
16 瓶巴洛羅紅酒中,評比挑選出最佳酒莊並
呈獻給亞洲讀者,然對我們來說,巴洛羅紅
酒所以為「義大利之紅酒之王」,其尊貴優
雅的多層次口感,無以言喻,只能由每個人
親自開瓶並品嘗。

我確信筱雯的〈Barolo Library 餐桌上的義大利酒王〉活動將會非常的成功,且
我認為這只是個開端、且待未來更多巴洛羅紅酒在亞洲的發展。

Leading Judge / Federico Scazello

In my opinion, Barolo Library project is one of the most important work that
has never developed in Asia before. For this reason it's important to recognize to
Xiaowen that her determination was fundamental to develop her idea in a very
professional way. I followed with big interest the tastings in Italy in the participation
of important chefs and sommeliers in luxury locations. After a blind taste of 16
Barolo, we select the top ones and present to Aisa, yet for us the quality of Barolo
isn't to TELL but to open and TASTE yourself.

I'm sure this project will get a great success and I hope that this edition will be the
first for many others in the future.

5th Judge Tasting

— Piemonte, Italy —

第五場 / 巴洛羅評審會

主評人 **Leading judge**：Marco Antonucci
　　　　　　　　　　　　Amy Marcelle Bellotti
主　廚 **Chef**：Marc Lanteri

評審地點 LOCATION

一星米其林餐廳 / 北義葛林察內‧卡梧爾百年城堡
Castello di Grinzane Cavour/1-star Michelin Restaurant
義大利皮爾蒙特省葛林察內卡梧爾 Grinzane Cavour, Piemonte, Italy

餐廳的歷史

葛林察內・卡梧爾城堡是北義最具代表也最重要的城堡之一。上一位城堡主人是卡米羅公爵,他於 1832 年至 1849 年擔任葛林察內市總督時便居住於此,公爵同時也是義大利統一復興時期的重要建築師之一。目前城堡內依舊收藏著公爵的設計手稿、家具、總督公文等等文物。

The Castle once belonged to Count Camillo Benso of Cavour, who lived here from 1832 to 1849, being the Mayor of Grinzane. He was also one of the important architects of the Unification of Italy and in this castle, his furniture, manuscripts and mayoral tricolour sash are well-preserved.

廚介紹

・蘭特里主廚出生於阿爾卑斯山旁的法義邊境小鎮,他菜餚是擁有義大利精神的法國高端料理,他結合了法國普旺斯與北義皮爾蒙特省的兩大傳統傳統,並使用山林區與中海料理食材,共同激盪出充滿靈性與技巧的絕佳美味。

ef Marc Lanteri was born in Tende in the Maritime Alps. s French haute cuisine has deep Italian roots. He brings life an encounter between mountains and sea: a subtle mbination of Provence, France and Piedmont, Italy. Lanteri's isine is one of instinct and training combined.

筱雯推薦的〈不可錯過〉What you should not miss!

網址 website: www.castellogrinzane.com
地址 address: Via Castello, 5, 12060 Grinzane Cavour CN, Italy
電話 tel: +39 0173 262172
推薦菜餚 Xiaowen's Recommended Dish:
北義傳統小牛燉肉鮪魚醬 (Vitello Tonnato)、烤羊肩與鷹嘴豆法式薄餅佐北義特有品種黑橄欖與羅勒醬
(Grive di Agnello)

評審團
JUDGES

馬可・安東努齊

義大利美食教育中心常務監事、義大利品油專家協會理事長、橄欖油官方評審、資深美食記者

義大利建築設計師，同時也是品油專家、義大利官方指定橄欖油評審；著義大利橄欖油相關專業書籍，並時常於義大利電視與媒體教導品油；多年來於世界各國推廣義大利橄欖油文化。

Marco Antonucci

Managing Supervisor of CLUBalogue Academy, Food Journalist, and Panel leader of EVOO

First of all Architect, journalist specialized in food, professional taster of olive oil and international panel leader. For many years internationally engaged in spreading the culture of olive oil through seminars, courses, meetings, guides, articles and publications.

馬・蘭特里

義大利一星米其林主廚

擁有義大利精神的法國高端料理；結合山林區與地中海料理食材；師習於法國三星主廚 Alain Ducasse 與 Enzo Santin 等諸多米其林餐廳；於 2004 年在北義得到第一顆米其林星級肯定。

Marc Lanteri

1star Michelin chef, Ristorante Al Castello Grinzane Cavour, Piemonte

French Italian cuisine that has an individual and recognizable style. His professional training includes Alain Ducasse, Christian Morisset and Michel Rostang, followed by Enzo Santin, Annie Feolde and Paolo Teverini with his first Michelin Star in 2004 in Cuneo, Piedmont.

艾咪·瑪雀樂·貝蘿蒂

義大利伺酒師協會皮埃蒙特省伺酒師、城堡女主人
出生於美國克羅拉多州、義大利亞斯提廚藝學校取
得 Master Chef 認證、並擁有專業伺酒師認證；她
從 1998 年起便居住於義大利並協助她的米其林主
廚丈夫（馬克蘭特里）經營其餐廳。

Amy Marcelle Bellotti

Hostess and Wine Sommelier

Born in Colorado USA; obtained her Master Chef
certification in Costigliole d'Asti; a Sommelier
Professionalista. Since 1998, she assists her
husband and business partner, Marc Lanteri, in the
management of their restaurant.

維樂利歐·貝爾特密

義大利資深伺酒師

已於餐飲業打滾 40 年，既是品酒教授亦為專業經
理人。他這一生對葡萄酒的熱情，促使他以資深專
業伺酒師的身分任教於義大利、瑞士、以及蒙地卡
羅的重要餐廳。

Valerio Beltrami

Mastre and Wine sommelier

For 40 years as teacher of wine and professional
manager, his passion and love for wine has made
him Sommelier and grand "Maestro" of restaurant,
responsible for Italy, Switzerland and Monte Carlo.

亞歷山卓・蔻爾辛尼

義大利伺酒師協會皮埃蒙特省伺酒師

25 年餐飲經驗、現擔任米其林餐廳伺酒師；熟悉且熱愛義大利有機酒和自然酒，其技巧應用搭配於當季菜餚、賓主盡歡；對於北義葡萄酒擁有豐富的知識 (其中當然包括巴洛羅紅酒)。

Alessandro Corsini

Wine Sommelier

Wine sommelier at Al Castello Ristorante, with more than 25 years of work experience. His passion and respect for organic and biodinamic wine producers is transmitted in his suggested wine pairings; great knowledge of Piedmont's best wines, Barolo included.

山下由佳

日本餐飲視覺設計師

現居於義大利亞斯提，她的料理專長為義日混搭的原粹料理，目前致力於推動小農食材。

Yuca Yamashita

Food Stylist, Japan

Lives in Asti, Piemonte; specialized in Itaponese, fusion of Italian and Japanese cuisine. Now she is working for little producers.

大衛・寶培羅
義大利酒愛好者、義大利食品公司老闆
我是個無可救藥的航海夢想家！儘管寬廣開放的大自然令我徜徉無限，我的巧克力工廠依舊是我最大的熱情。我的巧克力如我人一般，是如此的簡單、善良、令你一個接著一個停不下來。

Davide Barbero
Wine lover and owner of food company
I am a hopeless dreamer who loves sea and sailing! Despite my passion for outdoors and open spaces, my chocolate factory is still THE place. My products say who I am, because they reveal some traits of my being: a simple and genuine goodness, taste after another.

16 支不同巴洛羅紅酒
16 BAROLO SELECTED

 16 支不同的巴洛羅紅酒 16 Barolo selected

◆ 酒莊與編號 Winery and No.　▲ 葡萄坡名 CRU　 評審會盲飲結果 Judge choices
● 酒名與年份 Wine/Vintage　■ 酒精濃度 Alcohol degree

<u>01</u> p.143
◆ Sebaste Mauro (06_2)
● Barolo PRAPÒ Docg 2011
▲ Prapò, Serrelunga d'Alba
■ 14.5%vol.

<u>05</u> p.170
◆ Fratelli Casetta (26)
● Barolo Case Nere Docg 2009
▲ Case Nere, La Morra
■ 14%vol.

<u>02</u> p.158
◆ Massolino (19_1)
● Barolo Classic Docg 2011
▲ Mix, Serralunga d'Alba
■ 14.5%vol.

<u>06</u> p.173
◆ Podere Rocche dei Manzoni (28_1)
● Barolo Bricco San Pietro Vigna d'la Roul Docg 2010
▲ Bricco San Pietro Vigna d'la Roul, Monforted'Alba
■ 14%vol.

<u>03</u> p.161
◆ Virna di Borgogno Virna (21)
● Barolo Cannubi Boschis Docg 2010
▲ Cannubi Boschis, Barolo
■ 14%vol.

<u>07</u> p.181
◆ Raineri Gianmatteo (34)
● Barolo MONSERRA Docg 2010
▲ Perno, Monforte d'Alba
■ 14.5%vol.

<u>04</u> p.165
◆ Sordo Giovanni (22_4)
● Barolo PARUSSI Docg 2011
▲ Parussi, Castiglione Falletto
■ 14.5%vol.

<u>08</u> p.185
◆ Gianni Ramello (37)
● Barolo ROCCHETTEVINO Docg 2010
▲ Rocchettevino, La Morra
■ 14%vol.

09 p.196 TOP1

◆ Ciabot Berton (44_2)
● Barolo ROCCHETTEVINO Docg 2009
▲ Rocchettevino, La Morra
■ 14.5%vol.

13 p.215

◆ Giovanni Rosso (66_2)
● Barolo SERRA Docg 2011
▲ Serra, Serralunga d'Alba
■ 14%vol.

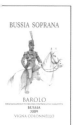

10 p.197 TOP2

◆ Bussia Soprana (45)
● Barolo BUSSIA Vigna Colonello Docg 2009
▲ BUSSIA Vigna Colonello, Monforte d'Alba
■ 14%vol.

14 p.218

◆ Luigi Pira (70_2)
● Barolo MARGHERIA Docg 2010
▲ Margheria, Serralunga d'Alba
■ 14.5%vol.

11 p.202

◆ G.D.varja (46_5)
● Barolo Albe Docg 2011
▲ Mix, Barolo
■ 15%vol.

15 p.222

◆ Giacomo Fenocchio (75_1)
● Barolo VILLERO Docg 2011
▲ Villero, Castiglione Falletto
■ 14.5%vol.

12 p.212

◆ Poderi Luigi Einaudi (62_1)
● Barolo Terlo Vigna Costa Grimaldi Docg 2010
▲ Terlo Costa Grimaldi, Barolo
■ 14%vol.

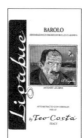

16 p.228

◆ Teo Costa (88_1)
● Barolo Ligabue Docg 2010
▲ Mix, La Morra
■ 14.5%vol.

給讀者的話
JUDGES COMMEND

主評人 / 馬可‧安東努齊

我認識筱雯已經有一段時間，而她對於研究食物與義大利文化的堅持、往往令我為之著迷而佩服。今天我們受邀做為評審，也是因為我們如同筱雯，對於義大利這塊土地擁有著豐富的知識及熱情，因此我們一同品評、那曾經生長於土地上的葡萄、經由釀酒師的雙手轉化成為一瓶瓶美酒的巴洛羅紅酒。如同完整的人生的旅程，這一瓶瓶巴洛羅經過「人」的雙手成為美酒，也要經過「人」們與其好友的共同開瓶分享其完美口感，方才完成其旅程的使命。正如我們在義大利與筱雯共享、正如我們現在與書前的你(我們的讀者)分享，我們希望你也會在這世界某一角落、與你自己的朋友們、開一瓶巴洛羅，共享美妙時光！

Leading Judge / Marco Antonucci

I know Xiaowen for some time now and she has always fascinated me of her continuous research from passion and love for food and Italian culture. Like Xiaowen, all of us have great knowledge and passion for our land for, the food and the wine which is clear to be the reason why we are here. We tasted the Barolo that are transformed from grape of each vineyard to excellent wine through hand of men. The journey of the bottles doesn't stop until being tasted by "men" again who share the excellence quality and joy with friends, like what we do in Italy with Xiaowen and to all of you who are reading this book right now. We hope you will do the same with your friends in different corner of the world.

主評人 / 艾咪・瑪雀樂・貝蘿蒂

品嘗第一杯巴洛羅
我們還是一群原先不認識的美食餐飲專業人士；
第三杯
我們是互相諮詢、互相尊重彼此的同事；
第十三杯
我們是熟識且輕鬆談話的朋友；
第十六杯
我們已經是好朋友！

筱雯事先挑選 16 瓶皆來自歷史悠久酒莊之知名
巴洛羅紅酒，各瓶皆代表不同單一葡萄坡之風
土條件，這使得評比結果變得更加困難。經過
一整天的評審，除了挑選出最佳巴洛羅推薦給世界另一端的華人讀者外，我覺
得、這世上沒有比「美酒加好友」更令人享受「即刻生活」的經驗！

Leading Judge / Amy Marcelle Bellotti

We were unfamiliar food and wine professionals with the first glass, consulting collegues by the 3rd glass, easy acquaintance at the 13th glass and good friends by the 16th! Xiaowen's selection of fine Barolo wines made the judging very difficult. Today we not only selected the top Barolo to recommend to the other side of the world, but through fine wine and good friends we stopped time for a quiet instant allowing us to enjoy life's pleasures !

6th Judge Tasting

— Milano, Italy —

第六場 / 巴洛羅評審會

主評人 **Leading judge**：Antonio Di Mora
主　　廚 **Chef**：Pietro Leemann

評審地點 LOCATION

一星米其林餐廳 Restaurante JOIA/ 1-star Michelin Restaurant
義大利倫巴第省米蘭 Milano, Lombardia, Italy

餐廳的歷史

「JOIA」來自於英文的 Joy 及義大利文的 Gioia，兩者皆是歡愉的意思，概念是讓用餐成為一件愉悅且能發現自我的事。主廚的綠色素食料理呈現西方式禪學；東方跟西方對於料理中的五感有著不同的詮釋，但不變的是、享受美食能為我們帶來內心的歡愉。

The name JOIA comes from Joy (English) and Gioia (Italian), both meaning HAPPINESS- the main concept and the spirit of dishes here. The vegetarian cuisine of Pietro Leemann represents his philosophy of united culture of the west and of the east. Though they are different in interpretation of experience from the 5 senses, the happiness and pleasure that one perfect dish brings is universal.

主廚介紹

皮耶托·里蒙被稱譽為義大利素食米其林教父，他的素食料理源自於義大利傳統原味然呈現獨創一格的「西方禪學」，極具創造力及蔬食原始色彩，以東方文化的禪意 (形狀) 融合進西式料理的變化 (義大利傳統原味)。

Pietro Leemann, the first Michelin chef of Italian vegetarian cuisine; his dishes are from Italian tradition taste that represents his philosophy in union of western and of the east and that embodies creativity and originality of each ingredients through different color and shape. He promotes diet of healthy and of green which naturally brings HAPPINESS as JOIA represents.

筱雯推薦的〈不可錯過〉What you should not miss!

網址 website: www. joia.it

地址 address: Via Panfilo Castaldi, 18, 20124 Milano, Italy

電話 tel: +39 0229 522124

推薦菜餚 Xiaowen's Recommended Dish:

九宮格的顏色盤 (Colori, gusti e consistenze)、主廚松露米蘭燉飯 (L'ombelico del mondo)、榛果巧克力塔佐香脆牛奶冰淇淋及覆盆莓優格 (Ricordo)

評審團
JUDGES

皮耶托·里蒙
義大利一星米其林主廚

義大利素食米其林教父,他的素食料理源自於義大
利傳統原味且呈現獨創一格的「西方禪學」;他著
有諸多素食哲學書籍並極力推廣綠色飲食。

Pietro Leemann
1star Michelin Chef, JOIA, Milano

The first vegetarian Michelin chef; his dishes are
from the taste of Italian traditions and represent
his philosophy in union of western and of the east;
He has many books on vegetarian cuisine and his
philosophy. He dedicates himself to promoting healthy
and green diet.

馬可·廷堤
義大利一星米其林副主廚

道地的托斯卡尼人,一口流利的義大利文與其方言;
當他談到美酒美食時,他便停不下來地侃侃而談。
在廚房裡,他能夠快速地將每一道米其林菜餚完美
呈現;目前為一星米其林副主廚。

Marco Tinti
Sous chef, 1star Michelin Chef, JOIA, Milano

Born in Tuscany, speaks Italian and his local dialect.
When he talks about food and wine, he can't stop.
In the kitchen, he is the historical mind of JOIA who
can easily organize the work. For 5 years he is the
sous chef in JOIA.

大衛・拉瑞世
羅馬五星級飯店主廚

出生於北義另一大紅酒產區，自然而然地、他有著自己獨到的葡萄酒味蕾；師廚米其林主廚里蒙，將其綠色蔬食哲學從米蘭帶到羅馬。

Davide Larise
Chef, Hotel Raphael, Roma

Born in the production area of Amarone, one of the important wine of Italy, naturally he has his own taste in wine. Following his work with Chef Leemann, he brings JOIA's vegetarian philosophy from Milan to Roma.

卡羅・索曼以尼
瑞士官方電視台記者

他的一生不停地漂泊於世界每一個角落，一年當中有六個月追隨美食美酒的腳步、在世界旅行；美酒美食文化的研究，就是他的人生。他的專長在紀錄各國食物傳統，其影片在歐洲得到許多大獎肯定。

Carlo Someini
Journalist, Swiss TV

In his whole life, he travels around the world. He and his camera follows the footprint of good food and wine in each countries and make documentary of its traditions. This is his life style and his works has been recognized with awards in Europe.

薇薇安‧拉波托莎
瑞士電視明星主廚

出身於瑞士義大利區，她熱愛廚房中的變化並推崇「一口文化」，也就是歐洲盛行的手指小食；著有〈義大利手指小食：用手吃的 200 道食譜〉、〈手指小食 ： 一口品嘗的 140 道食譜〉等書籍。

Viviana Lapertosa
Chef, Swiss

Born in Switzerland, speaks Italian as mother tongue. She loves everything about kitchen and is passionated for finger food. Her publication includes "Finger food all'italiana. Oltre 200 ricette da mangiare con le mani", "Finger food. 140 ricette da mangiare in un boccone" and others.

安東尼‧帝‧摩拉
義大利伺酒師協會隆巴地亞省伺酒師

與里蒙主廚共同工作長達 16 年，負責米其林餐廳的一切營運，工作中當然包括葡萄酒與菜餚的搭配與遊戲；他熱愛他的工作以及全世界的葡萄酒。

Antonio Di Mora
Wine Sommelier and restaurant manager

He has been working with Chef Leemann for 16 years and is in charge of everything of JOIA, which includes of course, the matching between food and wine. For him, wine is the best "game" to play with. He loves his work and the world of wine.

德拉・比歐塔・賈克蒙
米其林餐廳專業伺員

年輕且具有強烈求知慾；對於巴洛羅紅酒情有獨鍾。
師習米其林主廚，日後可望獨當一面管理餐廳。

Della Briotta Giacomo
Chef de Rang

He is a young boy but with a strong desire to learn, especially toward Barolo wine. He learns how to manage a restaurant in high standard and will become the manager one day.

陳伯昌
台灣食品公司老闆

食品進口商，也是個大吃貨，進口上千種食物，目前在台北擁有數家餐廳。他正在進入義大利葡萄酒的世界，是評審團當中唯一的葡萄酒「素人」。

Frank Chen
Owner of Taiwan Food Company

Own a food company and a foody himself. He imports different kind of food and has several restaurants in Taipei such as Japanese Teppanyaki. He doesn't know much of Italian wine, yet he is on his way.

16 支不同巴洛羅紅酒
16 BAROLO SELECTED

16 支不同的巴洛羅紅酒 16 Barolo selected

◆ 酒莊與編號 Winery and No.　　▲ 葡萄坡名 CRU　　 評審會盲飲結果 Judge choices
● 酒名與年份 Wine/Vintage　　■ 酒精濃度 Alcohol degree

__01__ p.133

◆ Castello di verduno (03_2)
● Barolo MONVIGLIERO Docg 2008
▲ Monvigliero, Verduno
■ 14.5%vol.

__05__ p.166

◆ Sordo Giovanni (22_5)
● Barolo RAVERA Docg 2011
▲ Ravera, Novello
■ 15%vol.

__02__ p.137

◆ Cascina Adelaide (04_4)
● Barolo CANNUBI Docg 2010
▲ Cannubi, Barolo
■ 14.5%vol.

__06__ p.167

◆ Sordo Giovanni (22_6)
● Barolo ROCCHE DI CASTIGLIONE Docg 2011
▲ Rocche di Castiglione, Castiglione Falletto
■ 14.5%vol.

__03__ p.146

◆ Michele Chiarlo (09)
● Barolo CEREQUIO Docg 2010
▲ Cerequio, La Morra
■ 14%vol.

__07__ p.174

◆ Podere Rocche dei Manzoni (28_2)
● Barolo PERNO Vigna Cappella di S. Stefano Docg 2010
▲ Perno, Monforte d'Alba
■ 14%vol.

__04__ p.163

◆ Sordo Giovanni (22_2)
● Barolo MONVIGLIERO Docg 2011
▲ Monvigliero, Verduno
■ 15%vol.

__08__ p.177

◆ Cavallotto Tenuta Bricco Boschis (31_1)
● Barolo BRICCO BOSCHIS VIGNA SAN GUISEPPE Docg 2008
▲ Bricco Boschis, Castiglione Falletto
■ 14%vol.

09 p.198 TOP1

◆ Luigi Baudana-G.D.varja (46_1)
● Barolo BRICCO DELLE VIOLE Docg 2011
▲ Bricco Delle Viole, Barolo
■ 14.5%vol.

13 p.216

◆ Poderi Colla (69)
● Barolo BUSSIA Dardi le Rose Docg 2010
▲ Bussia Dardi le Rose, Monforte d'Alba
■ 14%vol.

10 p.199 TOP5

◆ Luigi Baudana-G.D.varja (46_2)
● Barolo RAVERA Docg 2011
▲ Ravera, Novello
■ 14.5%vol.

14 p.219

◆ Luigi Pira (70_3)
● Barolo MARENCA Docg 2010
▲ Marenca, Serralunga d'Alba
■ 14.5%vol.

11 p.201 TOP4

◆ Luigi Baudana-G.D.varja (46_4)
● Barolo CERRETTA Docg 2011
▲ Cerretta, Serralunga d'Alba
■ 14.5%vol.

15 p.223

◆ Giacomo Fenocchio (75_2)
● Barolo BUSSIA Docg 2011
▲ Bussia, Monforte d'Alba
■ 15%vol.

12 p.213

◆ Poderi Luigi Einaudi (62_2)
● Barolo CANNUBI Docg 2011
▲ Cannubi, Barolo
■ 14.5%vol.

16 p.227

◆ Abbona (85_3)
● Barolo CERVIANO 2009
▲ Cerviano, Novello
■ 14%vol.

給 讀 者 的 話
JUDGES COMMEND

主評人 / 安東尼・帝・摩拉

義大利巴洛羅紅酒以及亞瑪蘿納紅酒是我的最愛。巴洛羅來自皮爾蒙特省的朗格樂，也是義大利最重要的紅酒產區。如果酒莊能承沿傳統並在每個環節小心工作，便能得到最高品質、那完美的巴洛羅紅酒。有些人說義大利的巴洛羅好比法國的勃根第，原因在於朗格樂獨特的風土環境（土壤與氣候），我非常同意這個觀點。

這一場評審會的每一刻都非常有趣，因為所有評審一開始是如此安靜地品嘗、而後開始一起紛紛討論；每一位評審對於每一款的巴洛羅都有自己的想法和意見，我甚至可以使用「熱烈」二字來形容現場的討論氣氛。我個人偏愛「法威亞（酒莊名）」和「維士農城堡（酒莊名）」這兩家酒莊，因為我喜愛他們獨特風土環境給予葡萄酒優雅的態度與特殊的生命力。我要謝謝筱雯並且代表所有評審推薦在亞洲的讀者們、來參加這樣特殊且美好的經驗。

Leading Judge / Antonio Di Mora

Barolo and Amarone are my two favorite wines. Barolo is from Langhe of Piemonte which is a very important area of wine. If the winery follows its tradition and creates good work, we can have the wines of the highest level, of the excellent. Someone says Barolo is our（Italian）Borgogna because it is unique in terror and in climate which I could not agree more.

The tasting was a moment of great interests where we tasted the wines in silence but after we discussed together. It was even intense because each degustarori espreso has their ideas and impressions about different Barolo. Personally I love the Barolo of Vajra and Castle di Verduno because I find them very elegant and territorial. I want to thank Xiaowen for the nice tasting Barolo and on behalf of all the judges, I hope to recommend our readers this unique and wonderful experience.

索 引
I n d e x

風土條件與陳釀法定規格
/ 百瓶巴洛羅的詳細介紹 /
Technical Sheet of Each Barolo

酒莊索引 (按字母排序)
Contact of Each Winery

Aurelio Settimo (**01**)

A garnet red wine with plenty of structure and an intense bouquet
showing fruity and floral notes with hints of spices.
The flavour is full-bodied, with fine tannins making it smooth and well-balanced.
Bottled in APRIL 2013 – LOT LC13 – Alc. 14% by Vol.

Type	**Barolo DOCG**
Cru'	**ROCCHE DELL'ANNUNZIATA**
Town	**La Morra, Frazione Annunziata**
Altitude	**270 mt.**
Solar Exposure	**South – South-West**
Vine	**Nebbiolo - Lampia**
Production Area	**Piemonte – La Morra – Frazione Annunziata**
Stump hectare	**4,500 - 5,000**
Breeding kind	**Guyot**
Year	**2009**
Harvest	**Beginning October**
Improving	**36 months, 24 of which in oak casks**
Ageing	**15 -18 years**
Services Temperature	**18-20 ºC**
Wine making	**Traditional, ca. 15-20 days of maceration**
Colour	**Intense garnet red**
Gastronomic's Combining	**Game, roasts, braised meat, cheeses**
Kg of grapes for hectare	**8,000**
Grapes' Provenance	**La Morra – Frazione Annunziata**
Soil	**Calcareous**
Surface	**3,42 Ha.**
Total bottles produced	**6,413 (750 ML) - LOT C13**
Alcoholic Strength	**Alc. 14% by Vol.**

2 GLASSES　　**17/20**　　　**92/100**　　　　**90/100**　　　**17/20 (90/100)**　**91/100**

Bersano (02)

BADARINA BAROLO DOCG

A single vineyard from the "Badarina" estate in Serralunga d'Alba, rich, powerfull and luxurious, it will withstand prolonged ageing period and it will keep its freshness.

It represents one of the noblest wines of Bersano range, which boasts a tradition of at least half century in specific sector.

VARIETY NEBBIOLO – HARVEST END OF OCTOBER – VINIFICATION IN STAINLESS STEEL – REFINEMENT 30 MONTHS IN OAK CASKS, 6 MONTHS IN FRENCH BARRIQUES, 6 MONTHS IN THE BOTTLE

Intense ruby red with deep garnet hues. Strong scent bouquet of backberry and ripe red fruit with pleasant spicy and balsamic notes. Game, roasts and strongly flavoured cheeses.

Excellent longevity, over 10 years.

75 cl., 150 cl.

AWARDS

Wine Spectator Usa 91 Points

Duemila Vini Guide

Associazione Italiana Sommelier 4 Grapes

Vini Buoni d'Italia Touring 3 Stars

Decanter Special Barolo Tasting 4 Stars Highly Recommended

Luca Maroni "Annuario dei Migliori Vini" IP87

I Vini d'Italia Guide L'Espresso 3 Bottiglie

I vini di Veronelli Guide 3 Stars

BERSANO

Castello di Verduno (03_1)

CASTELLO di VERDUNO
Barolo Docg 2009 Massara

Vineyard
The Massara vineyard faces south-east and lies at 265 m. a.s.l. in the village of Verduno, on parcel 137p of land registry map sheet X (planted partly in 1988 and the remainder in 1992). The total area is 1.22,49 hectares, all planted with the Nebbiolo variety. The white soil is mainly chalky clay, known as the "Marne di Sant'Agata" (30% sand – 55% clay – 15% limestone). From this kind of soil we obtain very elegant and fragrant wines, with a great potential ageing.

Picking and maceration
The grapes are handpicked , sorting only the best bunches and placing them into a 20 kilos baskets. The harvest took place on October 6th 2009. The maceration in "tino" (open top wooden vat) lasted for 40 days, with regular punching down of the cap and, at the end of the alcoholic fermentation, we used the traditional technique of the "submerged cap" to continue with a static maceration.

Ageing
The wine aged for 32 months in large Slavonian oak casks and 2 months in steel tanks, followed by 21 months in the bottle prior to its release on September 2014.

Bottling
The wine was bottled, without any fining nor filtration, on October 1st 2012, with lot no.L12-10. The 2009 production has been of 5294 Albeisa bottles, 330 half bottles, 203 magnums and 17 Jeroboams.

Tasting notes
Clear garnet red. A complex and warm blend of herbs, blackberries, mint, sweet liquorice and ripe fruits.
Full-bodied and flavoursome with firm tannins and a long tasty finish.

Analytical Data
Alcool   14.50% Vol.
Acidity   5.8 g/l

Awards
93 points – Wine Enthusiast
17/20 – Jancis Robinson
4 stars - Decanter

Vintage 2009

The 2009 winegrowing season will be remembered for a winter characterized by particularly heavy snowfall, which guaranteed considerable water reserves that turned about to be extremely useful considering the lack of rainfall for most of the summer.
Climatic conditions were good during September and October and the fine weather favoured the grape harvest, both for early and late ripening grape varieties.

Castello di Verduno (03_2)

Barolo Docg 2008 Monvigliero Riserva

Vineyard
The Monvigliero vineyard is located in the commune of Verduno, at 280 m. a.s.l. on parcels 39 and 40 of land registry map sheet X (both planted in 1967). The total area planted with vines – all of the Nebbiolo variety - is 0.3 hectares. The vineyard lies towards the top of a hill, facing due south on white soil consisting of chalky clay , known as the "Marne di Sant'Agata" (30% sand – 55% clay – 15% limestone) From this kind of soil we obtain wines elegant, perfumed, with a good potential ageing.

Picking and maceration
The grapes were harvested by hand on October 10th 2008, with careful grading. The maceration in steel lasted 35 days, with regular punching down of the cap and, at the end of the alcoholic fermentation, we used the traditional technique of the "submerged cap" to continue with a static maceration.

Ageing
The wine was aged for 33 months in oak and 2 months in steel, followed by 54 months in bottles lying down prior to its release on April 2014.

Bottling
Bottling took place without filtration on August 9th 2011, with lot no. L11-12. A total of 2226 Albeisa bottles holding 0.75 l. were produced in the 2008 vintage.

Tasting notes
Clear garnet red. A complex and warm blend of herbs, jam, blackberries, mint, sweet liquorice and ripe fruits.
Full-bodied and flavoursome with firm tannins and a long tasty finish.

Analytical data
Alcohol: 14,5% by Vol.
Total acidity: 5,86 g/l

Particularities
The age of the vineyard (37 years) and the characteristics of the soil produce a wine each year with a marked structure, very elegant style, and a long cellar life.

Vintage 2008

The year 2008 will be remembered for a growing trend particularly climate that has characterized in two distinct phases. The first is characterized by heavy rainfall in spring, with peaks higher in the month of May. Fortunately, since the climate was more favorable in the last part of the season, it allowed to recover a year that, in early August, it seemed almost compromised. This providential weather allowed Nebbiolo grapes to reach ideal ripening, the wines are excellent!

Cascina Adelaide (04_1)

BAROLO "FOSSATI"

Fossati, ready to seduce you. Wine with all powerful of Barolo wines and all scents and elegance tipically of La Morra area. The last of our single-vineyard Barolo. The icing on the cake.

WINE CLASSIFICATION: Barolo DOCG.

GRAPE VARIETY: 100% Nebbiolo (5 different clones)

PLACE OF ORIGIN

This privately-owned vineyard at "Fossati" in La Morra. Exposure: South South-East.
Altitude: about 400 m
Soil: tortonian, limestone and blue clay.

THE VINEYARD

The vines are Guyot grown with the countertrellis system of planting, and medium-short pruning. The vineyard has an average density of 5000
vine-stocks per hectare, and the production is reduced by the cluster thinning method.

VINIFICATION

The grapes are collected when they are ripened and laid into small 20 kg perforated boxes. A strict cluster selection is made. The grapes are taken to the wine cellar and crushed within a few hours from the harvest. The fermentation takes place at the controlled temperature of about 32° C with repeated and dalicate pumping over. The marc macerationis long, 350 hours on average . The new wine is then put in wooden barrels, where the subsequent spontaneous malolacticfermentation takes place.

FINING

For 24 months in oak barrels; then, the wine is fined for 6 months in bottle.

TASTING NOTES

Dark, red ruby colour unmistakable character of La Morra's Barolo.
In the glass, great richness in black ripened fruit as plums and mulberry, then balsamic and spice eruption drawing together with the tannic layers to create a wonderful wine. It will delight us for decades.

FOOD MATCHES

"Barolo is the king of wines", but also a "king's wine" suitable for the cuisine of strong but elegant tastes, red meat dishes, game, truffle-based recipes, marbled or mature cheese.

ABOUT SERVING THIS WINE

The bottle, to be kept horizontally while stored, must be served at a temperature of 18° / 20° C. Cannubi is a great and elegant wine. Ageing: 15 - 20 years from the harvest onwards.

ANALITYCAL INFORMATION

Alcohol content: 14%/vol.
Residual sugars: 2 g/l
Total acidity: 6 g/l
Dry residue: 31 g/l

Cascina Adelaide (04_2)

BAROLO *Pernanno*

BAROLO " PERNANNO"

We are looking forward to taste this Barolo !
Finally it's here the first trial of the Pernanno vineyard grew on the Elvetian soil.
White clay burned from the sun which allowed to the nebbiolo grapes to reveal: unmatched smoothness, and trasparency in the plot. This distinctive quality can be found only on this side of Barolo area.

WINE CLASSIFICATION: Barolo DOCG.

GRAPE VARIETY: 100% Nebbiolo.

PLACE OF ORIGIN

Privately –owned vineyards extends for 0,4Ha, is located at "Pernanno" (very well known Castiglione Falletto crus) 2500 vine-stocks , planted around the 1980.
Exposure: sud- sud est
Altitude: 270m on the sea level
Soil: Elveziano(white-brown clay)

THE VINEYARD

The vines are Guyot grown with the countertrellis system of planting, and medium-short pruning. The vineyard has an average density of 5000 vine-stocks per hectare, and the production is reduced to about 1kg per vine-stock by the cluster thinning method.

VINIFICATION

The grapes are collected when they are fully ripened and laid into small 20kg perforated boxes. A strict cluster selection is made. The grapes are taken to the wine cellar where they stay for 12 hours in a "drying room" lowering the dampness above 5%.
The crushing operations, are very careful and soft to preserve the skin quality and not to have the grapestalk in the must. The fermentation takes place thank to indigenous yeasts at the controlled temperature of about 32°C with repeated and delicate pumping over. The marc maceration is long, 350 hours on average. The new wine is then put in wooden barrels, where the subsequent spontaneous malolactic fermentation takes place.

FINING

For 24 months in oak barrels; then, the wine is fined for 6 months in bottle.

TASTING NOTES

Red garnet colour. Incipit of minth and officinal herbs followed by roses bouquet keep alive thanks to an inner energy that seems neverending. Tannins,sweet and very thin in grain create harmony and complexity tipical of a very special cru.

FOOD MATCHES

Wonderful with based white truffle dishes and great companion of "Tagliolini and Agnolotti" .

ABOUT SERVING THIS WINE

The bottle, to be kept horizontally while stored, must be served at a temperature of 18- 20°C . Cannubi is a great and elegant wine. Ageing: 10 – 15 years from the harvest onwards.

ANALITYCAL INFORMATION

Alcohol content: 14,5%/vol.
Residual sugars: 2 g/l
Total acidity: 6 g/l
Dry residue: 30 g/l

www.cascinaadelaide.com

Cascina Adelaide (04_3)

BAROLO "PREDA"

Preda is a real "strong point" for us: a Barolo that makes every tasting special. It results from an accurate selection and care from the vineyard up to the wine cellar.

WINE CLASSIFICATION: Barolo DOCG.

GRAPE VARIETY: 100% Nebbiolo (2 clones).

PLACE OF ORIGIN

Historic privately-owned vineyard, located in the Preda area, municipality of Barolo. Exposure: West. Soil: tortonian, calcareous-clayey. Altitude: 270-300 m

THE VINEYARD

The vines are Guyot grown with the countertrellis system of planting, and medium-short pruning. The vineyard has an average density of 5,700 vine-stocks per hectare, and the production is reduced by the cluster thinning method.

VINIFICATION

The grapes are collected when they are fully ripened and laid into small 20kg perforated boxes. A strict cluster selection is made. The grapes are taken to the wine cellar where they stay for 12 hours in a "drying room" lowering the dampness above 5% the crushing operations are very careful and soft, to preserve the skin quality and not to have the grapestalk in the must. The fermentation takes place thank to indigenous yeasts at the controlled temperature of about 30°C with repeated and delicate pumping over. The marc maceration is long, 350 hours on average. The new wine is then put in wooden barrels, where the subsequent spontaneous malolactic fermentation takes place.

FINING

For 24 months in oak barrels; then, the wine is fined for 6 months in bottle.

TASTING NOTES

Preda shows a bold red ruby colour. This great wine amazes for it's intensity with plum jam and mulberry then developed menthol and iodate tones. The structure is huge in the mouth full bodied and licorice and ripened re fruit aromas are really clear in the glass.

FOOD MATCHES

"Barolo is the king of wines", but also a "king's wine" suitable for the cuisine of strong but elegant tastes, red meat dishes, game, truffle-based recipes, marbled or mature cheese.

ABOUT SERVING THIS WINE

The bottle, to be kept horizontally while stored, must be served at a temperature of 18-20 °C. Ageing: 10 – 15 years from the harvest onwards.

ANALITYCAL INFORMATION

Alcohol content: 14%/vol.
Residual sugars: 2,5 g/l
Total acidity: 6,1 g/l
Dry residue: 31 g/l

Cascina Adelaide (04_4)

 BAROLO *Cannubi*

BAROLO " CANNUBI"

In Barolo, the Barolo wine. One of the most prestigious crus in the area makes this wine unique. It becomes the ambassador of the quality and excellence of a whole territory.

WINE CLASSIFICATION: Barolo DOCG.

GRAPE VARIETY: 100% Nebbiolo, selezione massale.

PLACE OF ORIGIN

This privately-owned vineyard extends for 0.5 Ha , and is located at "Cannubi" (one of the most prestigious Barolo "crus"). 2,500 vine-stocks planted in 1980.
Exposure: South-South-East.
Tortonian - Elvet soil.

THE VINEYARD

The vines are Guyot grown with the countertrellis system of planting, and medium-short pruning. The vineyard has an average density of 5000 vine-stocks per hectare, and the production is reduced to about 1kg per vine-stock by the cluster thinning method.

VINIFICATION

The grapes are collected when they are fully ripened and laid into small 20kg perforated boxes. A strict cluster selection is made. The grapes are taken to the wine cellar where they stay for 12 hours in a "drying room" lowering the dampness above 5%.
The crushing operations, are very careful and soft to preserve the skin quality and not to have the grapestalk in the must. The fermentation takes place thank to indigenous yeasts at the controlled temperature of about 30°C with repeated and delicate pumping over. The marc maceration is long, 300 hours on average. The new wine is then put in wooden barrels, where the subsequent spontaneous malolactic fermentation takes place.

FINING

For 24 months in oak barrels; then, the wine is fined for 6 months in bottle.

TASTING NOTES

Our Cannubi shows delicate violet and plums so clear that seem in the glass a minerality and spicy aromas thanks to the inner energy of this unbelievable cru comes out like a flame.

FOOD MATCHES

"Barolo is the king of wines", but also a "wine for the king" suitable for the cuisine with strong but elegant tastes, red meat dishes, game, truffle-based recipes, marbled or mature cheese.

ABOUT SERVING THIS WINE

The bottle, to be kept horizontally while stored, must be served at a temperature of 18- 20°C . Cannubi is a great and elegant wine. Ageing: 10 – 15 years from the harvest onwards.

ANALITYCAL INFORMATION

Alcohol content: 14,5%/vol.
Residual sugars: 2 g/l
Total acidity: 6 g/l
Dry residue: 30 g/l

www.cascinaadelaide.com

Cascina Adelaide (04_5)

BAROLO *Baudana*

BAROLO BAUDANA
Historical single vineyards in Serralunga village, epitome of firmness and depth really unmatchable!

TASTING NOTES
Baudana shows a red ruby colour with a garnet hint. The nose is ethereal, magnificient, dense and lavish of yellow flower, white ripened peaches and dried apricot.
Silky and crunchy tannins, long lasting and very flavorful thanks to the minerality tipical of Serrlaunga.

WINE CLASSIFICATION: Barolo DOCG.

FOOD MATCHES
Risotto with porcini, lamb, seasoned cheese.

GRAPE VARIETY: 100% Nebbiolo, selezione massale

ABOUT SERVING THIS WINE
The bottle, to be kept horizontally while stored, must be served at a temperature of 18- 20°C . Cannubi is a great and elegant wine. Ageing: 10 – 15 years from the harvest onwards.

PLACE OF ORIGIN
Privately owned Vineyard of 0,7 ha in Serralunga village.
Exposure: south, Altitude: 300 m on the sea level
Soil: Elveziano

ANALITYCAL INFORMATION
Alcohol content: 14,5%/vol.
Residual sugars: 2 g/l
Total acidity: 6 g/l
Dry residue: 30 g/l

THE VINEYARD
The vines are Guyot grown with the countertrellis system of planting, and medium-short pruning. The vineyard has an average density of 5000 vine-stocks per hectare, and the production is reduced to about 1kg per vine-stock by the cluster thinning method.

VINIFICATION
The grapes are collected when they are fully ripened and laid into small 20kg perforated boxes. A strict cluster selection is made. The grapes are taken to the wine cellar where they stay for 12 hours in a "drying room" lowering the dampness above 5%.
The crushing operations, are very careful and soft to preserve the skin quality and not to have the grapestalk in the must. The fermentation takes place thank to indigenous yeasts at the controlled temperature of about 32°C with repeated and delicate pumping over. The marc maceration is long, 350 hours on average. The new wine is then put in wooden barrels, where the subsequent spontaneous malolactic fermentation takes place.

CASCIN**A**DELAIDE

BAROLO
DENOMINAZIONE
DI ORIGINE CONTROLLATA
E GARANTITA

Baudana

FINING
For 24 months in oak barrels; then, the wine is fined for 6 months in bottle.

CASCIN**A**DELAIDE
www.cascinaadelaide.com

rimaldi Bruna (05_1)

BRUNA GRIMALDI

BAROLO DOCG BADARINA

BAROLO: the king of wines. Robust, full-bodied, with intense perfumes, a long finish and a garnet red colour with orange reflections. It is proposed in three versions: Badarina, Camilla and Bricco Ambrogio.
Cru Badarina is located in Serralunga d'Alba, in the heart of the Barolo district, with a south/south-eastern exposure, at 420 m. a.s.l., the vines are about 25 years old.

BLEND
100% Nebbiolo

PRODUCTION ZONE
Serralunga d'Alba

SOIL
calcareous, of medium consistence

ALTITUDE
420 meters (1400 feet) above sea level

FERMENTATION
traditional, with 20/30 days of skin contact

AGING
24-30 months in tonneau and in large wooden barrels

BOTTLE AGING
8-10 months

COLOR
an intense and brilliant garnet with orange highlights

AROMA
ample and intense, with sensations of violets, pepper, raspberries, and dried roses along with the classic notes of tar

FLAVOR
robust, persistent, and elegant

ALCOL MEDIUM
14.5°

MATCHING WITH FOOD
with roasts, stews, game, flavorful cheese, chocolate

SERVING TEMPERATURE
in crystal stemware for fine red wine, at a temperature of 18° centigrade (64° Fahrenheit)

Grimaldi Bruna (05_2)

BRUNA GRIMALDI

BAROLO DOCG CAMILLA

BAROLO: the king of wines. Robust, full-bodied, with intense perfumes, a long finish and a garnet red colour with orange reflections. It is proposed in three versions: Badarina, Camilla and Bricco Ambrogio.
Cru Camilla is in Grinzane Cavour, with a south/south-western exposure, at 220-250 m. a.s.l.

BLEND
100% Nebbiolo

PRODUCTION ZONE
Grinzane Cavour

SOIL
calcareous, medium consistence

ALTITUDE
220-250 meters (725-825 feet) above sea level

FERMENTATION
traditional, with 15/20 days of skin contact

AGING
24-30 months in tonneau and large wooden barrels

BOTTLE AGING
approximately 8-10 months

COLOR
intense and brilliant garnet with orange highlights

AROMA
ample and intense, with notes of violets, pepper, raspberries, and dried roses along wiht th classic tar

FLAVOR
full, persistent, and elegant

ALCOL MEDIUM
14°

MATCHING WITH FOOD
roasts, stews, game, flavorsome cheese, chocolate

SERVING TEMPERATURE
at 18° centigrade (64° Fahrenheit) in crystal stemware for important red wines

Grimaldi Bruna (05_3)

BRUNA GRIMALDI

BAROLO DOCG BRICCO AMBROGIO

BAROLO: the king of wines. Robust, full-bodied, with intense perfumes, a long finish and a garnet red colour with orange reflections. It is proposed in three versions: Badarina, Camilla and Bricco Ambrogio.
Cru Bricco Ambrogio is in Roddi, with a south-south-eastern exposure, at 280-320 m. a.s.l.

BLEND
100% Nebbiolo

PRODUCTION ZONE
Roddi d'Alba

SOIL
medium-textured calcareous clay

FERMENTATION
traditional, with 20/25 days of skin contact

AGING
24 months in large wooden barrels

BOTTLE AGING
8/10 months

COLOR
an intense garnet red, brilliant and with orange highlights

AROMA
ample and intense with notes of violets, pepper, raspberries, and dried roses along with the typical notes of tar

FLAVOR
full and firm, long and elegant

ALCOL MEDIUM
14.5°

MATCHING WITH FOOD
roasts, game, full-flavored cheese, chocolate

SERVING TEMPERATURE
18°centigrade (64° Fahrenheit) in crystal stemware for red wines

索引

Sebaste Mauro (06_1)

Barolo Trèsüri Docg

GRAPE VARIETY:
100% Nebbiolo.

ORIGINS:
Three different vineyards with excellent exposure to the sun, each one with its characteristic, find their highest exaltation joined together this coming to be of high quality Barolo.

TYPE OF SOIL:
Limestone.

HARVESTING TIME:
October.

VINIFICATION:
From 12 to 16 days in stainless steel tanks with temperature controlled automatic and computerized mixing and pressing.

AGING:
In French oak barrels from 1600 liters to 400 liters and a few carats for 36 months.

BOTTLING:
After the necessary aging in barrel at least 8 months prior to release.

COLOR:
Ruby red with garnet reflections.

NOSE:
Fine and elegant with beautiful shades of cherry, rose and sweet licorice.

PALATE:
Full and harmonious with soft and elegant tannins, long and persistent.
DEGREES:
13.5-14%

FOOD PAIRINGS: Red meats or cheeses of medium or long aging.

Sebaste Mauro (06_2)

Barolo Prapò Docg

GRAPE VARIETY:
100% Nebbiolo

ORIGINS:
The Prapò vineyard is a plot in the Commune of Serralunga d'Alba.

TYPE OF SOIL:
Calcareous and rich in iron

HARVESTING TIME:
October

VINIFICATION:
From 12 to 16 days in steel vats with carefully controlled temperature.

AGEING:
In 1600-liter French barrels, and in some small 400-liter barrels for 36 months.

BOTTLING:
After the necessary evolution in the cellar in barrels, at least 8 months before being marketed.

COLOR:
Ruby red, lively, and intense with slightly orange hues.

NOSE:
Ethereal nose and aromas of withered roses fail to suppress a slight fragrance of underbrush, which makes it extremely well-balanced and rounded.

PALATE:
Austere and impressive. With its excellent tannicity, it can have a very lengthy cellar life: 20/25 years. This Barolo reflects the most typical style of the wine's virtuosity and aristocracy.

DEGREES:
13,5-14%

FOOD PAIRINGS:
Could be savored alone, or with beef and some of the more subtle cheeses.

索引

Giribaldi (07)

Azienda Agricola Giribaldi
Rodello – Barolo (Piemonte) Italy

BAROLO D.O.C.G.
2010

GEOGRAPHICAL LOCATION
These grapes come from the vineyard Ravera in the community of Novello at a height 450 meters above sea level, the soil is marl - calcareous with a tendency to clayey, with a good permeability. The vineyard has a complete South exposure.

VINE AND PLANTING
The vineyard is composed of Nebbiolo (Michet, Lampia and Rosè) with Italian cloning material. The training system is Guyot with 5.000 plants per hectare.

TECHNOLOGICAL MENTIONS
The wine making process is in the traditional way, where for 18 days "rimontaggi and follatura" take place i.e. the must is pumped over the cap several times a day to avoid drying out, and also the cap is mechanically broken up. This happens at a temperature around 30°C to optimize extraction. The wine is then drawn off, and left in steel vats for 4 months, then is matured in oak casks for 2 years. Then up to today the wine is kept in the bottle in our underground cellar where there is a constant temperature of 16°C.

CHARACTERISTICS OF THE WINE
A ruby red colour with amaranth reflexes, an evident perception of good fruit appeals to the nose, floreal with hints of roses and cherries, dry and austere in the mouth, optimal ribbing with sweet tannins and sensations of tabacco.
An elegant wine with hints of spices that give pleasant sensations of freshness and softness.

AWARDS:
Barolo 2010: Decanter, Tom Maresca: 5/5 stars
Barolo 2010: 90 p. James Suckling
Barolo 2008: 91 points Wine Spectator
Barolo 2007: 90 points Wine Enthusiast
Decanter Magazine. London. May 2009 ***Three Stars for Giribaldi Barolo 2004
Wine Enthusiast (USA) December 2007:92 points
Decanter Magazine- London. *** RECOMMENDED Spicy, rich, appealing. Lovely balance, maturing beautifully. Rich and full- quite softy structured, but irresistibaly tasty. Lovely style. Up to 10 years.
Concorso Enologico "Acqui Terme": 25° ediz. Premio "Caristum" (best wine absolutely on all categories)
Concorso Internazionale Enologico "Estonian Wine Challenge" Estonia
Premio: Trofeo- "Trofee Winner" (prima premio assoluto)
Gewinner Der Verkostung Pro Wein. Germany: ** Barolo docg,
International Enological Wine Competition "Wine Enthusiast" USA: 92 points
International Wine & Spirits Competition- England: Silver medal
International Enological Wine Competition "Decanter Magazine" England.
Premio: **** 4 stelle per Barolo docg. "Lively, supple; excellent fruit."

Rivetto (08)

RIVETTO

Barolo DOCG del Comune di Serralunga d'Alba

GRAPE VARIETAL 100% Nebbiolo. A native Piedmont grape with a late harvest period, recognized for its quality since the ancient Roman epoch. The wine obtained is characterized by extraordinary refinement, exceptional ageing capacity and strong tannin content.

MUNICIPALITY Serralunga d'Alba

TOTAL EXTENSION 16,450 square metres

SELECTION IN THE VINEYARD In stages, carried out three times in the 45 days before the start of the harvest

HARVEST Manual

WINE MAKING Stalk removal, pressing and room temperature maceration for 2 days.Duration of maceration in total 20 days.Controlled temperature fermentation at 28 °C.

YEASTS Native

CLARIFICATION Fresh egg white

FILTRATION With inert material

AGEING 30 months in 30 hectolitre oak barrels from Slavonia

IN-BOTTLE REFINEMENT 10 months

SERVING TEMPERATURE 17°C

RECOMMENDED PAIRINGS Red meats and game

MANOCINO VINEYARD Sheet 9 Parcel 167/168/169/172

ALTITUDE 410 metres **EXPOSURE** East, North-East

TERRAIN Terrain of Tortonian origin, very clayey and calcareous. Terrain with sub-alkaline reaction, with strong magnesium presence. This vineyard is situated at a very elevated height which permits us to obtain Nebbiolo grapes of good acidity and with late maturation.

SERRA VINEYARD Sheet 9, Parcel 83

EXPOSURE South-East

TERRAIN Terrain of Tortonian origin, clayey-calcareous. Terrain with sub-alkaline reaction, with strong calcium presence and rich in magnesium with a fair presence of iron.

SAN BERNARDO VINEYARD Sheet 9 Parcel 84/85/4/3

ALTITUDE 355 metres **EXPOSURE** East

TERRAINTerrain of Tortonian origin, clayey-calcareous. Terrain with sub-alkaline reaction, with very strong calcium and magnesium presence.

索引

Michele Chiarlo (09)

Barolo Cerequio
Denominazione di origine controllata e garantita

Variety: Nebbiolo 100%
Vineyard: Cerequio, La Morra e Barolo
SurfaceArea: 3 hectares of 9
Rootstocks per hectare: 4800
Year of planting: 1972
Yield per rootstock: On average 1200 grams, equal to 55-58 quintals per hectare.
Soil composition: Calcareous Sant'Agata marl, basic PH, poor in organic substances, but rich in microelements, particularly magnesium and manganese.
Exposition: South, south - west
Average height a.s.l.: 320 metres.
Refinement: in 700 litre French oak barrels for 2 years. Afterward 15-16 months in bottle.

CHARACTERISTICS
Colour: Deep garnet with brilliant highlights and fine fluidity.
Aroma: Ample, generous, intense, with great complexity and balance, a symphony of sensations suggesting blackcurrant, apricot, cherry, mint leaves, gentian and spices.
Taste: All the aromas are confirmed on the palate, in a texture of rich full body, structure of great character, aristocratic tannins and an elegant, lingering finish.
DEVELOPMENT: Consume after 5-6 years, it ages well up to 20 years.
SERVINGTEMPERATURE: 17°-18° C (63°/ 65° F)
PAIRINGS: Robust red meat and game dishes, excellent with aged cheese.

Manzone Giovanni (**10_1**)

BAROLO CASTELLETTO 2009

Varietà di uva: 100% Nebbiolo
Origine: Castelletto, Monforte d'Alba, Langhe
Periodo di raccolta: prima- seconda decade di ottobre
Vinificazione: macerazione per 30 giorni a 28-31°C
Invecchiamento: in botti di Slavonia e tonneaux francesi per 36 mesi
Imbottigliamento: 40 mesi dopo la vendemmia

Bottiglie prodotte: N° 5.800
Tipologia di terreno: Calcareo argilloso
Altitudine: 340-400 m.s.l.m
Età delle vigne: 14 anni
Esposizione: est, sud est
Sistema di allevamento: Guyot semplice
Resa per ettaro in quintali: 55
Prima annata prodotta: 2004

Vita media: 20 anni
Conservazione: bottiglia coricata al riparo dagli sbalzi di temperatura e della luce
Temperatura di servizio suggerita: 16-18 °C

Esame Organolettico
Colore: rosso rubino con note granato di media intensità..
Profumo: Elegante ed intenso, con note di confettura di lampone, frutti di bosco e tabacco.
Sapore: caldo, piacevole, con ottimo equlibrio tannico e sapidità.
Gradi: 14%-14.5%
Abbinamento gastronomico: con selvaggina, pollame nobile e formaggi stagionati.

Manzone Giovanni (**10_2**)

BAROLO GRAMOLERE 2010

Varietà di uva: 100% Nebbiolo
Origine: Vigneto Gramolere, Monforte d'Alba, Langhe
Periodo di raccolta: prima- seconda decade di ottobre
Vinificazione: macerazione per 35 giorni a 28-31°C
Invecchiamento: in botti di Slavonia e tonneaux francesi per 36 mesi
Imbottigliamento: 40 mesi dopo la vendemmia

Tipologia di terreno: Calcareo argilloso
Altitudine: 340-400 m.s.l.m
Età delle vigne: 22-42 anni
Esposizione: sud, sud-ovest
Sistema di allevamento: Guyot semplice
Numero ceppi per ettaro: 5.000
Resa per ettaro in quintali: 65

Vita media: 20 anni
Conservazione: bottiglia coricata al riparo dagli sbalzi di temperatura e della luce
Temperatura di servizio suggerita: 16-18 °C

Esame Organolettico
Colore: rosso rubino con riflessi vivi e trama fitta.
Profumo: Intenso ed armonico. Buon frutto con ciliegia, lampone. Finale speziato e balsamico.
Sapore: Caldo, pieno, morbido, con ottimo equilibrio tannico e buona sapidità.
Gradi: 14-14,5 %
Abbinamento gastronomico:
con arrosti di carne rossa, selvaggina, formaggi stagionati

Manzone Giovanni (10_3)

BAROLO BRICAT 2009

Varietà di uva: 100% Nebbiolo
Origine: Vigneto Gramolere, Monforte d'Alba
Periodo di raccolta: prima- seconda decade di ottobre
Vinificazione: macerazione per 35 giorni a 28-31°C
Invecchiamento: in botti di Slavonia e tonneaux francesi per 36 mesi
Imbottigliamento: 40 mesi dopo la vendemmia

Bottiglie prodotte: N° 6.500
Tipologia di terreno: Calcareo argilloso
Altitudine: 340-400 m.s.l.m
Età delle vigne: 50 anni
Esposizione: sud, sud-ovest
Sistema di allevamento: Guyot semplice
Numero ceppi per ettaro: 5.000
Resa per ettaro in quintali: 55
Prima annata prodotta: 1994

Vita media: 20 anni
Conservazione: bottiglia coricata al riparo dagli sbalzi di temperatura e della luce
Temperatura di servizio suggerita: 16-18 °C

Esame Organolettico
Colore: rosso rubino con note granato.
Profumo: Ampio, vigoroso e persistente.
Note ricche di mora, mirtillo, noce e cioccolato.
Sapore: Complesso, elegante con tannino di grande effetto che avvolge il palato.
Gradi: 14-14,5%
Abbinamento gastronomico: brasati, arrosti di carne rossa, cibi tartufati e formaggi.

索引

BAROLO DOCG 2010

Barolo's wine born in these land but we dedicated love and meticulous work for the entire year. We start with the winter pruning of the vineyards ("guyot" method), followed by other jobs of allegation, cleaning and cutting of the vine until up to an accurate selection of the best bunches during harvest.

Then the process of the product's transformation begins: wine-making starts in the Barolo winery with pressing and the traditional fermentation of about thirty days. The entire philosophy of our Barolo Brunate is rigorously traditional, with fairly lengthy macerations followed by long maturations of about three years in 30 or 50-hectolitre oak Slavonian barrels and 12 months' bottleageing before release.

It can be enjoyed after a few months in the bottle, but will continue improving and developing its elegance during the years…. it is the true jewel of italian and international enology.

TASTING NOTES

Colur: Ruby red, very intensive red
Fragrance: Very pleasant and delicate, recalling wilted roses and notes liquorice and wood vanilla
Taste: Velvety, dry, rounded, rich and balanced in body and structure
Temperature: 18 – 19°C
Gastronomic's combining: Risottos, fondues, game, red meat and meat in general (roasts, grilled or skewered)
To drink: better in carafe

WINE MAKING

Harvest: Middle / End of October
Wine making: Classic method
Maturity: In bottles
Vintage: 2010
Ageing: In Slavonian wooden barrels
Period of maturity: 3 years in wood + 1 year approx in bottle
Bottle sized: 750 ml
Bottles produced (n°): 10.000 per year

VINEYARD

Area of production: La Morra – Brunate
Vine: Nebbiolo da Barolo 100%
Altitude: 350 m
Solar Exposure: South West
Breeding kind: spalliera / guyot method
Soil: Gravely clay
Year plant: 1970 with following renewals and 2000
Surface: 4 hectare
Kg of grapes for hectare: 8000 kg

CHEMICAL INFORMATIONS

Alcohol in vol (%): 14.60 % vol.
Total acidity: 5.30 g/L
Dry Extract: 26.7 g/
Total sugar: 0.4 g/L
Volative Acidity: 0,79 g/L
Total SO2: 59 mg/L

Sylla Sebaste_ (12)

SYLLA SEBASTE
BAROLO BUSSIA DOCG 2010

TASTING NOTES

This wine has a red garnet and shades of ruby red color which tends to turn to orange as the years go by. Its fragrance is made up of a bouquet of remarkable complexity, elegance and intensity, exceptional richness, harmony and completeness. From its fragrance of violet, rose and other freshly picked fruit it changes to the next evolution of the universe of spices. Its taste is pleasantly dry, full, robust, austere but velvety, harmonic and ethereal with a light aftertaste of liquorice and leather. Its lengthy persistence makesit great for accompanying red meat dishes, both national and internationa, wild game, braised meats and cheeses that have a strong and intense flavor.

BASIC INFORMATION

Product name: Barolo Bussia Docg

Vine type: Nebbiolo

Denomination: Barolo Docg

Classification: DOCG

Color: red

Typology: firm

Country/Region: Italy, Piedmont,

Vintage: 2006

Alcohol: 14.00

Aging: in oak barrels

Number of bottles produced: 5000

Silvano Bolmida (13)

BAROLO "BUSSIA" 2010

It comes from vineyards located in the higher part of Bussia of Monforte d'Alba, one of the most reknowned hill into the Barolo wine area. Very slope hill, 20 to 58 years old plant , at 420 mt on the sea level ; In total 1,7 ha in property, planted into a sandy-clay miocenic sea sediment, rich in flakes of grey marl. Very well sunny (facing to the south-west) and windy, easy to maintain safe cluster also in rainy vintage.

5.000 plant per hectare, cultivated with single guyot, not irrigated , with green manure fertilising based on "alfa –alfa" ; a carpet of this grass aid to avoid water erosion inside theese slope hills and aid to producing a particolar simbiosis with vine having so more quantity of parfums inside smaller size of clusters.

After many years of own experiment , starting from this year, controlling of fungal and insect maladies is made with natural vegetal exctract ; Against botritys (only few year i can had that kind of problem) i use saprophite ecological bacterial.

Green harvest is made in jun cutting in the half alls cluster ; in this way, from jun, the vine give to eat only to the berry that arrive into the cellar. Small crop per plant , around 1,2 kg aidm for more balance beetweeen sugar/acidity/total polifenol/ripe tannin.

Harvest is totally hand-made by my family, looking 13 analisyss to can arrive into the cellar with safe/ripe and weel balanced extract ; plastic cases are washed with hot water every times that we go to the Vineyard because-off a dirty case potentially produce a pollution of bad yeast and bacterial . Cluster have only destemming to can had at lot of whoole berry into the fermentation tank ; fermentation strat with a inoculation of selected-authocton fresh yeast . In this way i can had a big population that start in quickly way the fermentation, but this yeast population comes from Nebbiolo vineyards in the Langhe area.

Inside inox tank, fermentation with automatically punping over go on for one week, after there are a very long maceration post fermentation, on the submerged skyns, 65 days in total, and malolactic fermentation is made in quickly natural way in this moment ; the substrate is rich is nutrient substance, with no sulfite working, with skyns that capture (if there are) the secondary malolactic parfums.

Claudio Boggione (**14**)

VIGNA

Comune di produzione:	Barolo
Cru:	Brunate
Altitudine media:	300mt.s.l.m.
Esposizione:	Sud-Ovest
Terreno:	calcareo,argilloso con prevalenza Magnesio
Anni di impianto:	dal 1959 al 2010 con successivi Rinnovamenti
Superficie:	1.2 ha

VINO

Colore:	Rosso rubino intenso
Gradazione alcolica:	14,5% vol
Acidita' totale:	5,8 g.l.
Estratto secco:	30%
Affinamento e Invecchiamento:	botti di rovere di slavonia No barique
Bottiglie prodotte:	8000

Ettore Fontana _ Livia Fontana (15)

BAROLO DOCG VILLERO

Product name: Barolo DOCG Villero

Grape variety: 100% nebbiolo

Production zone: Castiglione Falletto

Vineyard: circa 1 hectare in Villero, 300-350 metres asl with southwest exposure

Harvest: hand-picked with rigorous grape selection into small crates towards the middle of October

Vinification: in temperature-controlled fermentation casks with frequent délestages and punching down.

Aging: in traditional oak barrels for ca. three years and a long period in the bottle in the air-conditioned wine cellar before release

Colour: intense ruby red with orange nuances

Tasting notes: Elegant and harmonious on the nose with persistent notes of fruit. Warm, full-bodied, well-balanced palate with long and intense finish

Food pairing: excellent with red meat, game and tangy, aged hard cheeses

Serving: in large glasses at 17-18°C (62–64°F)

Storage: horizontally in a cool, dry place with constant temperatures; ready to drink but can be cellared for 10/12 years

Bottle sizes: 75 cl. / 1,5 lt. / 3 lt.

Ilario Claudio (16)

Technical Data
Barolo "Sorano"

Technical Data:

Production area: **Serralunga (CN)**
Grape: **Nebbiolo 100%**
Age of vines: **10 anni**
Training system: **Guyot**
Density: **4000-5000 vines per hectare (Ha)**
Harvest: **early October**
Average yield per hectare (Ha): **60 ql**
Vinification: **in red with thermo rotofermentatori**
Ageing: **24 months in french barrique, 12 months in large barrels 30Hl and 2 years in bottle**

Analysis:

Alcohol: **14.50 % vol.**
Sugars: **2.80 g/l**
Total Acidity: **5.70**
PH: **3.40**

Organoleptic features:

Colour: **ruby red with garnet**
Scent: **typical, full, intense ripe fruit**
Taste: **Dry, warm, soft, tannic and persistent**
Duty Temperature (°): **16-18°**
Recommended pairing: **red meats and cheeses**

索引

Ca' Rome (18_1)

CA' ROME' BAROLO
RAPET

Vineyard area (estate-owned): 1 hectare (c. 2.4 acres)

Altitude: 310-371 meters (1,017-1,217 feet) above sea level

Exposure: Southwest   Soil: 53% clay, 28% silt, 19% sand, pH 8.00

Year of planting: 1972   Varieties: 100% Nebbiolo Michet and Lampia

Density: 5,000 vines per hectare (c. 2,025 per acre)

Crop yield: 1.3 kilos per vine

Vinification:

Traditional floating-cap fermentation, with 1-week-long submerged cap maceration; malolactic lasts 1 month, in oak. The wine is subsequently aged 2 years in wood, partly in 25-hl. barrels and partly in barriques (225 liters). Before release, it is bottle-aged for a few more months.

Tasting Notes:

Color: Garnet with orange reflections.

Bouquet: Ample, elegant and well balanced, with characteristic notes of dried roses and underbrush.

Palate: Full body and perfect balance of components, fruit, tannin and acidity in ideal synergy, lush flavors and pleasing, long finish.

Cellar Life:

In top vintages, 15-25 years.

a' Rome (18_2)

CA' ROME' BAROLO
VIGNA CERRETTA

Vineyard area (estate-owned): 1.7 hectares (4.2 acres)
This is a single vineyard within the Serralunga d'Alba township. Soil is
geologically marine sedimentary originating in the Middle and Upper Miocene
periods of the Tertiary era, and is principally compact, impermeable clay and
calcium carbonate marl of whitish and bluish hues. The subzone's heavy terrain
is conducive to robust, full-bodied and long-living reds with high alcohol
content. Regular soil analyses have shown the overall silty, clayey texture of the
land, typical of the Langhe area. Trace elements, indispensable for healthy,
quality grapes, are abundant.
Altitude: 310-371 meters (1,017-1,217 feet) above sea level
Exposure: Southwest   Year of planting: 1961
Varieties: 100% Nebbiolo Michet and Lampia
Density: 5,000 vines per hectare (c. 2,025 per acre)
Crop yield: 1.2 kilos per vine

Vinification:
Harvest (manual) takes place at the end of October. Temperature-controlled
fermentation lasts 3 weeks: the first 5-6 days, temperature is max. 30° C (86° F),
followed by 2½ weeks at max. 26° C (78.8° F). The wine is subsequently aged 2
years in oak, partly in 25-hl. barrels and partly in barriques (225 liters). After
élevage in wood, it is assembled in stainless steel, where it sojourns for a brief
period previous to 1 more year of bottle age before release.

Tasting Notes:
Color: Deep garnet, with coral reflections.
Bouquet: Ample, elegant and ethereal, very rich and fragrant, shows a wealth of
aromas reminiscent of dried fruit, roses, underbrush, tobacco, bay leaves and vanilla.
Palate: Full body, luscious flavors and a very long finish closing on almond
notes.

Cellar Life:
15-30 years, depending on vintage.

Massolino (19_1)

Barolo

DENOMINAZIONE DI ORIGINE CONTROLLATA
E GARANTITA

Prodotto con uve Nebbiolo provenienti da vigneti siti nel territorio di Serralunga d'Alba.

Altitudine: tra 320 e 360 m s.l.m.

Superficie totale: 6 ettari.

Tipologia del terreno: calcareo in generale con variazioni anche consistenti da zona a zona.

Sistema d'allevamento e densità d'impianto: Guyot tradizionale; da 5.000 a 6.000 viti per ettaro.

Resa per ettaro: 65 quintali.

Età media delle viti: da 10 a 55 anni in relazione al vigneto.

Vendemmia: manuale, effettuata nella seconda metà di ottobre.

Prima annata di produzione: 1911.

Totale bottiglie prodotte: 38.000 da 0,75 lt e 500 da 1,5 lt circa.

Gradazione Alcolica: 13,5-14% Vol., in relazione all'annata.

Vinificazione e invecchiamento: Barolo tradizionale con fermentazione e macerazione di 15 giorni circa a temperature varianti tra i 31 e i 33° C; l'invecchiamento avviene in botti di rovere per la durata minima di 30 mesi. Segue poco più di un anno di affinamento in bottiglia in appositi locali freschi e bui.

Note: Il Barolo DOCG classico riveste un ruolo di grande spicco per la nostra azienda: con questo vino desideriamo proporre una bottiglia di altissimo livello!

Caratteristiche di degustazione

Alla vista: rosso granato con intensità variabile in relazione all'annata.

Al naso: le uve provengono da diverse sottozone di Serralunga e, proprio per questo motivo, conferiscono una gamma di profumi molto ampia che può andare dalle accattivanti note speziate a quelle più dolci, floreali e fruttate.

Al palato: molteplici le sensazioni, ci troviamo di fronte ad un vino corposo, classico e ben strutturato che non teme l'invecchiamento e rappresenta in modo egregio il carattere importante delle nostre terre. È consigliabile caraffarlo e berlo a temperatura di 18-20°C. Si esprime al meglio se abbinato alle carni rosse, in particolare selvaggina e cacciagione, e a piatti tartufati. Ottimo anche su primi piatti di pasta all'uovo con sughi di carne e risotti come anche con formaggi vaccini e caprini di media stagionatura.

Serralunga d'Alba

MASSOLINO

Massolino (19_2)

Barolo

DENOMINAZIONE DI ORIGINE CONTROLLATA
E GARANTITA

Prodotto con uve Nebbiolo provenienti dal vigneto Parussi (foglio n. 2, parcelle n. 342, 108, 268) in Castiglione Falletto.

Altitudine: 300 m s.l.m.

Superficie totale: 1,3 ettari.

Tipologia del terreno: calcareo argilloso sabbioso.

Sistema di allevamento e densità di impianto: Guyot tradizionale con circa 5.000 viti per ettaro.

Resa per ettaro: 45 quintali.

Età media delle viti: 40 anni.

Vendemmia: manuale, effettuata nella seconda metà di ottobre.

Prima annata di produzione: 2007.

Totale bottiglie prodotte: da 4.000 a 5.000 da 0,75 lt in relazione all'annata.

Gradazione alcolica: 13,5-14,5% Vol. in relazione all'annata.

Vinificazione e invecchiamento: Barolo tradizionale con fermentazione e macerazione di 15-20 giorni ad una temperatura di 31-33° C; invecchiato in botti di rovere per circa 30 mesi e affinato in bottiglia per minimo un anno in appositi locali freschi e bui.

Note: l'eccellente esposizione sud-est/sud-ovest del vigneto, localizzato sulla sommità della collina in sottozona Parussi, garantisce la produzione di uve di straordinaria qualità.

Il terreno, molto calcareo, regala al vino un carattere deciso e austero, con tannini assai robusti.

Caratteristiche di degustazione

Alla vista: rosso granato intenso.

Al naso: profumo etereo ed avvolgente con intense e persistenti note di spezie dolci, sandalo, tabacco e cuoio.

Al palato: strutturato e potente, evidenzia, nei primi anni di vita, un'austerità notevole che si lima con il passare del tempo. I tannini molto fitti e robusti lo rendono certamente longevo. Il lunghissimo finale evidenzia il carattere di Castiglione Falletto.

È consigliabile decantarlo e servirlo ad una temperatura di 18-20° C.

Il Parussi è il primo Barolo della storia aziendale nato fuori dai confini "serralunghesi", in un altro vocalissimo terreno: quello di Castiglione Falletto. La vinificazione e l'affinamento tradizionale che condivide con i "fratelli" di Serralunga d'Alba esaltano la nota capacità dell'uva Nebbiolo di regalare espressioni diverse al variare, seppur minimo, del microclima e del terreno in cui viene coltivata.

Perfetto in abbinamento con i ricchi piatti della cucina piemontese, carni stufate e formaggi di media e lunga stagionatura.

Serralunga d'Alba

MASSOLINO

Bric Cenciurio (20)

BAROLO
Denominazione di Origine Controllata e Garantita
"Coste di Rose"

VINE VARIETY:	Nebbiolo divided in three clones: Michet, Lampia e Nebbiolo Rosè.
SOIL:	Under the surface rich in organic matter, to a depth of 50/100 cm we find an alternation of blue and grey strata of clay (named "MARNE DI S. AGATA FOSSILI"). In these sedimentary formations the vines sink their roots with difficulty in order to absorb the microelements which will be characteristic of this wine.
VINEYARDS:	Located within the commune of Barolo in the area that borders on Monforte d'Alba, in the "sottozona": "Coste di Rose". The soil in this area help the production of some pronounced young wines which have the capacity to mature, to improve and to reveal the elegance of this important Barolo wine. Exposure: South, South - East. Altitude: 320 metres above sea level. Age of th vineyard: 30 - 40 years. Training system: arbour; pruning system: Guyot. Strum/hectare: 4000 - 4500. Production: about 1 Kg./stump.
YIELD/HECTARE:	4000 - 5000 kg. of grapes.
GRAPE HARVEST:	Second 10 days of October.
VINIFICATION PROCESS:	After the crushing, the grapes stay one day at a temperature of 5-7 °c, then the fermentation and the maceration start and continue for 25-30 days. Just after the drawing process, the wine is refined for 2 years in large Slavonian oak cask (2500 liters). During the last period of the ageing the wine is carried out in stainless steel tanks for 6 months and then for 10 months in bottles.

ALCOHOLIC CONTENT: 13.5 - 14% vol.

BOTTLES PER YEAR: 3000 - 4000

irna di Borgogno Virna **(21)**

BAROLO CANNUBI BOSCHIS
D.o.c.g.

TECHNICAL INFORMATION

VINEYARD:

Vine: Nebbiolo, under variety Lampia.

Grapes provenance: Barolo "cru" Cannubi Boschis

VINEYARDS:

South-west facing with Guyot pruning (9.10 buds / vine)

N° Vines / hectare: 4,500

Yield / hectare: 50 q / Ha

Plant year: 1970

WINEMAKING:

Soft pressing of grapes, maceration with grape skins for around 10-12 days at a maximum temperature of 32°C.

After the malolactic fermentation in spring 50% the wine is decanted in to Tonneau (500 Litre) in which matures for 18 months and 50% in to oaken barrels from Slavonian with a capacity of 50-60 Hl in which the Barolo matures for 6-12 months.

After the clarification, the wine is bottled without filtration and stays in the bottles for one year before packaging and consignment. The best years became RISERVA after five years.

Alcoholic strength	: 14% vol.
Total acidity	: 5,00-6.00 g/l.
Dry extract	: 25-28 g/l.
Residual sugar	: < 3 g/l

ORGANOLEPTIC CHARACTERISTICS, PRESERVATION:

With a ruby-red colour, the Barolo has a rich bouquet which gradually recalls the scents of roses flows, truffles and wood spice. The palate is at first elegant and refined, then is begins to gain in complexity with air. The long flavours predict a prominent future ahead.

Sordo Giovanni (22_1)

Superficie: 3.8500 HA.
Lavori nel vigneto: allevamento viti col sistema Guyot. Inerbimento controllato, concimazioni organiche. Diradamento grappoli a luglio-agosto. Ampelopatie trattate con zolfo e solfato di rame.
Lavori in cantina: pigiatura e diraspatura, fermentazione tumultuosa a temperatura controllata, macerazione a cappello sommerso per tre-quattro settimane secondo tradizione, cura della fermentazione malolattica, maturazione e affinamento in botti di legno tradizionali.

Colore splendido, questo rosso granato intenso e profondo con la presenza positiva di luminosi riflessi rubini che promettono una lunga vita evolutiva di questo vino. Perfettamente brillante. Un evidente sviluppo di archetti e lacrime che scorrono fluide lungo le pareti di cristallo del calice sono indice di una grande generosità, struttura e forza alcolica.

Bouquet austero, nobile, importante, molto ampio e persistente. Ottima la franchezza.
Possiamo percepire tra le sensazioni robuste ma non invadenti del rovere, il profumo di frutta matura e la fragranza dei fiori secchi di montagna. Una leggerissima presenza di resina mette in risalto toni di spezie dolci e di foglie di pesco.

All'assaggio, dopo la prima sensazione di calore intenso, si evidenzia la presenza di tannini buoni che tendono al dolce. Sapidità molto spiccata tipica della terra. Si confermano, molto ampliati, i profumi e le sensazioni che avevamo percepito al naso.

Una grande struttura, armonia ed equilibrio uniti ad una lunga persistenza aromatica e complessità sono sicuri presupposti di una gloriosa longevità.

Un millesimo che si ricorderà per la sua eccezionalità.

Temp 20 °C.

BAROLO
SÖRI GABUTTI
Denominazione di Origine Controllata e Garantita

Surface area: 3.8500 HA.
In the vineyard: Guyot vine-training system; controlled grass regeneration, organic manuring; thinning of bunches in July-August; vine diseases treated with sulfur and copper sulfate.
In the winery: crushing and destemming, tumultuous fermentation at a controlled temperature, traditional submerged cap maceration for three-four weeks, malolactic fermentation, ageing in traditional wooden casks.

A magnificent color: deep, intense garnet red, with bright ruby highlights promising a long evolution. Perfectly brilliant. The legs that flow freely down the inside of the glass are a clear indication of this wine's power, structure and alcohol.

On the nose it shows itself to be a rich, thoroughbred wine with a long, ample finish. Excellent cleanliness. Noticeable among the robust, but unobtrusive sensations are oak, the aroma of ripe fruit, and the fragrance of dried mountain flowers. A very slight hint of resin brings out overtones of sweet spices and peach leaves.

An intense, warm entry on the mouth is followed by nice, sweetish tannins and a very marked earthy flavor. The aromas and sensations that showed on the nose are confirmed in more depth.

Great structure, harmony and balance, combined with a long aromatic finish and complexity, promise a glorious future.

A vintage which will long be remembered for its outstanding quality.

Sordo Giovanni (22_2)

Superficie: 1,36 HA

I **terreni** di quest'area sono costituiti da marne bianche fini abbastanza sciolte e ricche di sali minerali.
La sua altitudine è di circa 280 metri sul livello del mare.

Caratteristiche organolettiche:

Vino con un bellissimo **colore** rosso granato con riflessi rubini di invitante aspetto.
Archetti e lacrime ampie e persistenti danno risalto alla sua perfetta limpidezza.

Bouquet elegante, intenso, ampio e persistente. Si possono percepire facilmente i sentori della rosa appassita, del sottobosco, tabacco dolce e spezie.

Al **palato** denuncia una grande struttura, sensazioni di calore intense e piacevoli. Fresco nella sua giusta sapidità, interessante per i suoi tannini che volgono al dolce con leggeri accenni di caffè, cioccolato, frutta matura e spezie.
Vino di grande equilibrio capace di emozionare e soddisfare il consumatore più esigente.

Lunga la persistenza aromatica intensa.

Temp 20 °C

BAROLO
MONVIGLIERO
Denominazione di Origine Controllata e Garantita

Surface area: 1.36 HA

The **soil** here is composed of fairly loose, fine white marl with a high content of mineral salts.

Its **altitude** is around 280 metres above sea level.

Tasting notes:

Beautiful garnet **red** with inviting ruby nuances. Ample, lingering legs highlight perfect clarity.

Elegant, intense **bouquet** which is ample and long. Withered roses, underbrush, sweet tobacco and spices easily identifiable.

Great structure on the **palate**, with appealing intense, warm notes. Fresh flavour showing interesting sweetish tannins, with slight hints of coffee, chocolate, ripe fruit and spices. Wine with great balance, capable of exciting and satisfying the most demanding of palates.

Long and intense aromatic finish.

Sordo Giovanni (22_3)

Esposizione: Sud-Est
Altitudine sui 350 metri sul livello del mare.
Terreno tufaceo con strati di terra rossa, di non facile lavorazione. Il sottosuolo è costituito prevalentemente da sassi e roccia detritica cementata più o meno tenacemente da componenti sabbiosi e argilla.
Superficie: 5.4400 HA.
Lavori nel vigneto: allevamento viti col sistema Guyot. Inerbimento controllato, concimazioni organiche. Diradamento grappoli a luglio-agosto. Ampelopatie trattate con zolfo e solfato di rame.
Lavori in cantina: pigiatura e diraspatura, fermentazione tumultuosa a temperatura controllata, macerazione a cappello sommerso per tre-quattro settimane secondo tradizione, cura della fermentazione malolattica, maturazione e affinamento in botti di legno tradizionali.

Straordinario colore rosso rubino intenso con ampi riflessi granata che denotano una giovinezza non ancora sfiorata dagli effetti dei primi anni di maturazione. Vino trasparente e brillante con una formazione generosa ed evidente di archetti e lacrime che scorrono copiose lungo le pareti di cristallo del calice: questo per la sua spiccata forza alcolica e quantità di sostanze estrattive.

Il bouquet è franco, pulito, intenso, ampio, composito. Racconta di sentori delicati di frutta matura, di cacao, leggera vaniglia; di presenza di legni nobili, menta e spezie dolci.

In bocca, proviamo una prima sensazione di calore intenso e piacevole quasi dolce, nonostante la presenza importante e marcata di tannini maturi che si legano bene alla spiccata sapidità che completa la forza del vino. Ottimo il lento percorso evolutivo che promette e garantisce una grande longevità e ulteriori trasformazioni, in positivo, delle pur già notevoli sensazioni olfattive e gustative.

Vino di grande struttura, molto equilibrato e giustamente armonico.
Una lunga persistenza aromatica intensa gli conferisce grande razza e nobiltà.

Un vino che passerà alla storia.

Temp 20 °C.

BAROLO
CERETTA DI PERNO
Denominazione di Origine Controllata e Garantita

Facing: south-east
Altitude: approx. 350 meters above sea level.
Soil: tufaceous, with layers of terra rossa which is not easy to work. The subsoil is mainly formed of stones and clastic rock, held together by sand and clay.
Surface area: 5.4400 HA.
In the vineyard: Guyot vine-training system; controlled grass regeneration, organic manuring; thinning of bunches in July-August; vine diseases treated with sulfur and copper sulfate.
In the winery: crushing and destemming, tumultuous fermentation at a controlled temperature, traditional submerged cap maceration for three-four weeks, malolactic fermentation, ageing in traditional wooden casks.

Extraordinary shade of deep ruby red, with plenty of garnet nuances showing a youth that has barely been touched so far by the effects of several years of ageing.
A bright, transparent wine packed with alcohol and extractive substances, forming an abundance of clear legs that run freely down the glass.

The complex bouquet is clean and straightforward, hinting delicately at ripe fruit, cocoa and vanilla; noble wood, mint and sweet spices also showing.
Lovely warm entry on the mouth: almost sweet, despite the nice merging of marked ripe tannins with the unmistakable lusciousness that rounds off the strength of this wine. A long life and further, positive transformations of what are already remarkable sensations on the nose and palate are promised by its excellent slow evolution.

A wine of great structure, very rounded and with just the right balance. A long, intense aromatic finish puts the final touch to a wine of great breeding.

A wine which will go down in history.

Sordo Giovanni (22_4)

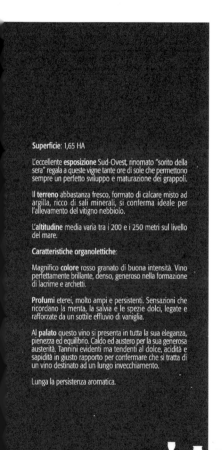

Superficie: 1,65 HA

L'eccellente **esposizione** Sud-Ovest, rinomato "sorito della sera" regala a queste vigne tante ore di sole che permettono sempre un perfetto sviluppo e maturazione dei grappoli.

Il **terreno** abbastanza fresco, formato di calcare misto ad argilla, ricco di sali minerali, si conferma ideale per l'allevamento del vitigno nebbiolo.

L'**altitudine** media varia tra i 200 e i 250 metri sul livello del mare.

Caratteristiche organolettiche:

Magnifico **colore** rosso granato di buona intensità. Vino perfettamente brillante, denso, generoso nella formazione di lacrime e archetti.

Profumi eterei, molto ampi e persistenti. Sensazioni che ricordano la menta, la salvia e le spezie dolci, legate e rafforzate da un sottile effluvio di vaniglia.

Al **palato** questo vino si presenta in tutta la sua eleganza, pienezza ed equilibrio. Caldo ed austero per la sua generosa austerità. Tannini evidenti ma tendenti al dolce, acidità e sapidità in giusto rapporto per confermare che si tratta di un vino destinato ad un lungo invecchiamento.

Lunga la persistenza aromatica.

Temp 20 °C.

BAROLO
PARUSSI
Denominazione di Origine Controllata e Garantita

Surface area: 1.65 HA

*Its excellent south-westerly **position**, which exposes it to the warmth of the sun for long hours in the afternoon and evening, allows for perfect development and ripening of the bunches.*

*The fairly cool **soil** composed of lime mixed with clay is packed with mineral salts, and is ideal for growing the nebbiolo variety.*

*The average **altitude** varies between 200 and 250 metres above sea level.*

Tasting notes:

*Magnificent garnet **red** showing good intensity. Perfectly bright, dense wine, producing plenty of legs.*

*Very long, expansive, ethereal **nose**. Sensations reminiscent of mint, sage and sweet spices, woven together and reinforced by a subtle aroma of vanilla.*

*On the **palate** this wine exudes elegance, body and balance. Warm and powerful, it clearly shows sweetish tannins, with just the right balance between acidity and flavour confirming its potential for lengthy ageing.*

Long aromatic finish.

Sordo Giovanni (22_5)

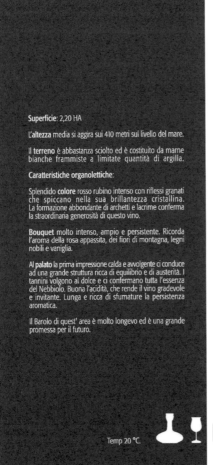

Superficie: 2,20 HA

L'altezza media si aggira sui 410 metri sul livello del mare.

Il **terreno** è abbastanza sciolto ed è costituito da marne bianche frammiste a limitate quantità di argilla.

Caratteristiche organolettiche:

Splendido **colore** rosso rubino intenso con riflessi granati che spiccano nella sua brillantezza cristallina. La formazione abbondante di archetti e lacrime conferma la straordinaria generosità di questo vino.

Bouquet molto intenso, ampio e persistente. Ricorda l'aroma della rosa appassita, dei fiori di montagna, legni nobili e vaniglia.

Al **palato** la prima impressione calda e avvolgente ci conduce ad una grande struttura ricca di equilibrio e di austerità. I tannini volgono al dolce e ci confermano tutta l'essenza del Nebbiolo. Buona l'acidità, che rende il vino gradevole e invitante. Lunga e ricca di sfumature la persistenza aromatica.

Il Barolo di quest' area è molto longevo ed è una grande promessa per il futuro.

Temp 20 °C

BAROLO
RAVERA
Denominazione di Origine Controllata e Garantita

Surface area: 2.20 HA

The **average** altitude is around 410 metres above sea level.

Tasting notes:

Splendid deep **ruby-red**, with garnet highlights that stand out on its crystal clearness.
The formation of plenty of legs confirms the extraordinary power of this wine.

Very intense, long, complex **bouquet**, conjuring up withered roses, mountain flowers, noble wood and vanilla.

Starts out warm and pervasive on the **palate**, leading into a big structure packed with balance and body. The sweetish tannins confirm all the essence of Nebbiolo. Good acidity gives the wine an appealing, inviting feel. Long aromatic finish full of nuances.

The Barolo from this area is very long-lasting, promising a great future.

Sordo Giovanni (22_6)

Superficie: 0.7600 HA.
Lavori nel vigneto: allevamento viti col sistema Guyot. Inerbimento controllato, concimazioni organiche. Diradamento grappoli a luglio-agosto. Ampelopatie trattate con zolfo e solfato di rame.
Lavori in cantina: pigiatura e diraspatura, fermentazione tumultuosa a temperatura controllata, macerazione a cappello sommerso per tre-quattro settimane, cura della fermentazione malolattica, maturazione e affinamento in botti di legno tradizionali.

Questo vino ci impressiona per la tonalità, l'intensità e la straordinaria bellezza del suo colore. Uno splendido insieme di granati e rubini che illuminano il cristallo del calice. Vino perfettamente brillante, che lascia tracce vinose dense sulle pareti di cristallo. La sua forza alcolica si esprime in una abbondante formazione di archetti e lacrime.

Ci sorprende per l'intensità, la ricchezza e la grande persistenza del suo bouquet. Le caratteristiche varietali del vitigno sono impreziosite ed ampliate da sensazioni nette ed eleganti che ci ricordano i frutti maturi delle ciliege nere, dei mirtilli e di confettura di pesche. Nuance floreali "fieno di montagna" liquirizia e tabacco cedendo il passo a quei nobili profumi terziari che si sono formati, in un primo tempo nelle botti di rovere da 15/20 hl. e, successivamente nell'ambiente ridotto della bottiglia. Vaniglia, sentori di sottobosco e di legni nobili completano le sensazioni olfattive.

Al palato abbiamo un inizio caldo e avvolgente. Tannini evidenti ma morbidi ed una buona presenza acida esaltano la sua struttura di vino possente, ricco, adatto a lunghi anni di affinamento. Negli aromi di bocca ritroviamo, ancora più intense, tutte le sensazioni che avevamo recepito al naso. Assai lunga la persistenza aromatica intensa. In conclusione ci troviamo, in questo momento, di fronte ad un vino austero e generoso, importante. Un vino che fa prevedere una vita molto lunga. Una sicura promessa per il futuro.

Un vino prestigioso che regalerà ai suoi estimatori sensazioni e memorie che non si potranno dimenticare tanto facilmente.

Temp 20 °C

BAROLO
ROCCHE DI CASTIGLIONE
Denominazione di Origine Controllata e Garantita

Surface area: 0.7600 HA.
In the vineyard: Guyot vine-training system. Controlled grass regeneration, organic manuring. Thinning of bunches in July-August. Vine diseases treated with sulfur and copper sulfate.
In the winery: crushing and destemming, tumultuous fermentation at a controlled temperature, submerged cap maceration for three-four weeks, malolactic fermentation, ageing in traditional wooden casks.

You cannot help but be struck by the hues, intensity and extraordinary beauty of this wine's color. The glass is lit up by the splendid combination of garnet and ruby nuances in its perfect brilliance. Dense vinous traces are left on the walls of the glass, with plenty of legs demonstrating the strength of its alcohol.

The intensity, richness and tremendous length of its bouquet are quite remarkable. Its varietal qualities are made more precious still by elegant, clear-cut sensations reminiscent of ripe black cherries, bilberries and peach jam. Hints of flowers, mountain hay, licorice and tobacco then give way to the noble tertiary aromas which formed first in the 15/20 hl. oak casks, and then in the confines of the bottle. The nose is rounded off by overtones of vanilla, underbrush, and noble wood.

On the palate it starts out warm and mouth-filling. Evident – though soft – tannins and good acidity bring out all the structure of a big, powerful wine suited to lengthy ageing. All the sensations showing on the nose are even more intense in the mouth. Very long aromatic finish. In conclusion, at the moment this is a big, austere wine that looks like having a very long cellar life. Definitely a wine of great promise.

A prestigious wine which will treat its admirers with sensations that will not easily be forgotten.

Le Ginestre (24)

LE GINESTRE

Barolo "sotto castello"

Growing location: the vineyard lies in the village of Novello, on the highly-prized "sotto castello" (below the castle) cru facing south, south-east at an average altitude of 400 m asl. The fairly deep limestone soil has medium consistency with good drainage. The vines are planted with a density of 4,500 per hectare, and are trained on vertical trellises and pruned using the Guyot system.
Year of planting: 1999 e 2002

Wine-making process: vinification on the skins, with a submerged cap and daily pumping over of the must. The maceration is extended for several days following the alcoholic fermentation by steeping the cap in order to increase the extraction and stabilize the colour extracted; after the wine has been drawn off and racked several times, it undergoes malolactic fermentation in steel before being transferred to oak by the end of the year to mature for at least two years prior to bottling, which takes place a year before the wine is released.

Tasting notes: deep, bright ruby red with slight orange hues; expansive, clear-cut nose, with hints of violet, vanilla, cinnamon, sweet tobacco and liquorice; very structured, soft mouth-filling taste, which is lightly dry with an appealing, long, spicy finish. Its structure and wealth of polyphenols give this wine an exceptionally long cellar life, during which its tasting properties continue to develop.

Roberto Cagliero (25)

AZIENDA AGRICOLA
CAGLIERO

PRODUCTION SHEET
BAROLO DOCG RAVERA

- **production zone:** Village of Barolo – cru Ravera; Village of Novello – cru Ravera
- **vineyards name:** Strà – Marenda
- **vineyards area:** more than 2,5 hectares
- **exposure:** South, South/East
- **altitude:** at 320 m above sea level, on hill side in the favourable slope
- **soil:** of medium consistency, mainly clay on a blue marl layer from the Tertiary period, with sand outcrops
- **grapes variety:** Nebbiolo
- **grapes yield per hectare:** 6000 kg after heavy thinning of the bunches
- **wine yield per hectare:** approx. 4000 litres
- **grape harvesting season:** generally the first two weeks of October
- **vinification:** the grapes, carefully harvested by hand, are immediately brought to the winery into suitable small aired crates and then destalked and soft pressed. The must is fermented at a maximum temperature of 32°C, that is controlled only through careful and frequent pumps-over. The skins are macerated for at least 10 days and when the fermentation is totally completed, that is all the sugar has been transformed into alcohol, the wine is racked off and transferred into stainless steel containers and the vinification process is completed with the malolactic fermentation
- **maturation:** most of the wine is matured for more than 2 years in Slavonian oak casks, while the remaining part is matured for about 12 months in medium toasted French oak tonneaux. Then the two wines are blended together and complete the maturation several months prior bottling. This way, a quite excellent elegant final wine is obtained, featured by full body and outstanding characteristics, big but perfectly well balanced structure, with intense scent. It is particularly suitable for ageing
- **fining out:** a period of more than 6 months in bottle completes the production process prior the wine can be released for consumption

索引

Lt. 0,75 -1,50 - 3
depending on the vintage

BAROLO CASE NERE D.O.C.G.

Barolo.... a name that recalls many things:
A great area with eleven hillside towns that constitute
the birthplace of Barolo and a wine that comes from far back in the past
Produced before 1600s and appreciated by the Royal Family of Savoy
House, it is notoriously defined as **"the king of wines, the wine of kings"**,
Barolo is today among the best expression of Italian enology, a robust and
complex wine, majestic expression of Nebbiolo grape

Case Nere is a single estate vineyard in La Morra village, know for its
elegant and excellent wines
Grape varieties: Nebbiolo 100%
Vineyard location: medium-high hillside (300-400 m asl)
with south-ouest exposure, distric of La Morra village.
Miocene-Helvetian soil origin with limestone and grey marls

Vinification

The grapes harvest is made by hand in the first/second ten days
of October.
The fermentation is traditional, carried out for an average
of 10-15 days in stainless steel vats at controlled temperature
with floating cap and frequent pumping over of the must to
optimize the extraction of the polyphenolic substances

Ageing

Minimum three years in large oak barrels from Slavonia and never
less than 6 months in the bottle.

Tasting Notes:

The colour is garnet red that takes slightly orange tone over time
Intense bouquet with withered roses and wild berries fruits overtones.
Full and harmonious on the palate, rich and persistent aromas that open
little by little. Complex wine, able to give unique sensations.
Great structure for long ageing.

Food pairing:

Excellent with meat dishes rich in flavors like braised beef, roast, lamb
and game.
The pleasure of tasting Barolo wine is not necessarily associates with the
delights of the food, but it is a sign of friendship and hospitality.
Serving Temperature: 18°-20° C

Dosio Vigneti (**27_1**)

Barolo D.o.c.g. Fossati
2009

Grapes: Nebbiolo

Vineyard:
Production zone: municipality of Barolo
Altitude: 330 m. above sea level
Soil composition: blue marls
Exposure: East – Southeast
Average yield: 60 quintals/hectare
Training system: Guyot
Harvest period: first ten days of October

Cellar:
Vinification: stemmer-crusher, maceration in temperature
controlled tanks at 25–26 °C with daily pumpover for 18–20
days and successive malolactic fermentation
Ageing: 40 months in 25-hl Slavonian oak barrels and then 9
months in glass before marketing

Tasting Notes:
Colour: garnet red with light orange tinges
Nose: violet and cherry fragrances that evolve into a light
note of chocolate, straw and leather
Taste: pleasingly dry and full thanks to its excellent balance
between the acidity and fine tannins of the Nebbiolo
Serving temperature: 16–18°C
Recommended pairings: red meats, game, braised meats,
aged cheeses

Alcohol volume: 14%

Number of bottles produced: 3,000 750 ml bottles

Duration of wine: 25–30 years

Dosio Vigneti (27_2)

Barolo
D.o.c.g. 2010

Grapes: Nebbiolo

Vineyard:
Production zone: municipality of La Morra
Altitude: 450–480 m. above sea level
Soil composition: blue marls
Exposure: South – Southeast
Average yield: 65 quintals/hectare
Training system: Guyot
Harvest period: first ten days of October

Cellar:
Vinification: stemmer-crusher, maceration in temperature
controlled tanks at 25–26 °C with daily pumpover for 15–18
days and successive malolactic fermentation
Ageing: 36 months in 25-hl Slavonian oak barrels and then 9
months in glass before marketing

Tasting notes:
Colour: garnet red
Nose: hints of cherry, rose and violet
Taste: excellent balance between acidity and tannins
Serving temperature: 16–18°C
Recommended pairings: red meats, game, braised meats,
aged cheeses

Alcohol volume: 14%

Number of bottles produced: 6,500 750 ml bottles

Longevity: 20-25 years

Podere Rocche dei Manzoni (28_1)

BAROLO Bricco San Pietro "Vigna d'la Roul"

APPELLATION	Appellation of controlled and guaranteed origin DOCG
VINTAGE	Yes
LOCATION OF THE VINEYARDS	Manzoni Soprani area- Podere Rocche dei Manzoni di Valentino in Monforte d'Alba – CN.
COORDINATES	
AGRICULTURE	Conventional - organic
IMPLANT	GUYOT 4000 vine-stocks per hectare
YELD	45-50 Q
GRAPE	100% Nebbiolo da Barolo.
WINEMAKING PROCESS	Accurate selection of the grapes. Fermentation for about 20-25 days in contact with the skin at a temperature between 25°C and 32 °C. Malolactic fermentation in oak barrels.
ALCOHOL CONTENT	14%
REFINING	It ages in oak barrels for 30 months and then it is put in egg-shaped containers for further 6 months. It refines still 12 months in bottle before sale.
LIFE OF THE WINE	Above 30 years.
TASTING NOTES COLOUR	Quite intense ruby red;
BOUQUET	Rich, wide and persistent, it reminds us of violets and goudront, with net and penetrating sensations;
TASTE	In the mouth it show an austere tannin of net punch. It is a well-breed wine, the most masculine wine of our production.
AVAILABLE IN	CL 75 /L 1,5
NOTES	Best served at a temperature of 18 °C. Best uncorked and decanted before serving.

Podere Rocche dei Manzoni (28_2)

BAROLO Perno "Vigna Cappella di S.Stefano"

APPELLATION	Appellation of controlled and guaranteed origin DOCG
VINTAGE	Yes
LOCATION OF THE VINEYARDS	S.Stefano Chapel – Perno area in Monforte d'Alba – CN.
COORDINATES	
AGRICULTURE	Conventional - biodynamic
IMPLANT	GUYOT 4000 vine-stocks per hectare
YELD	45-50 Q
GRAPE	100% Nebbiolo da Barolo.
WINEMAKING PROCESS	Accurate selection of the grapes. Fermentation at controlled temperature for about 20-25 days in contact with the skin. The must, rich in anthocyanins and coloured substances, is then separated from the skin and comes into barriques.
ALCOHOL CONTENT	14%
REFINING	It ages in oak barrels for 30 months and then it is put in egg-shaped containers for further 6 months. It refines still 12 months in bottle before sale.
LIFE OF THE WINE	Above 30 years.
TASTING NOTES COLOUR	Quite intense ruby red;
BOUQUET	Rich and persistent, wide and elegant, it reminds us of violets and marasca cherry;
TASTE	Complete and harmonic, it is a well-breed wine.
AVAILABLE IN	CL 75 /L 1,5 /L 3
NOTES	Best served at a temperature of 18 °C. Best uncorked and decanted before serving.

Podere Rocche dei Manzoni (29)

BAROLO "Pianpolvere Soprano Bussia"

APPELLATION		Appellation of controlled and guaranteed origin DOCG
VINTAGE		Yes
LOCATION OF THE VINEYARDS		Bussia, Monforte d'Alba – CN,
COORDINATES		
AGRICULTURE		Inspired by the biodynamic agricultural system
IMPLANT		GUYOT 4000 vine-stocks per hectare
YELD		Max 55 Q
GRAPE		100% Nebbiolo da Barolo.
WINEMAKING PROCESS		Accurate selection of the grapes. Fermentation for about 25-30 days in contact with the skin. Malolactic fermentation in small oak barrels.
ALCOHOL CONTENT		14,50%
REFINING		It ages in oak barrels for 48 months and then it is put in egg-shaped containers for further 12 months. It refines still 24 months in bottle before sale.
LIFE OF THE WINE		Above 30 years.
TASTING NOTES	COLOUR	Intense winy garnet red;
	BOUQUET	Great potentiality of variety: from red berries to tobacco and chocolate. Always in evolution;
	TASTE	Compact, of infinity power, but with great balance and elegance.
AVAILABLE IN		CL 75 /L 1,5/ 3L
NOTES		Best served at a temperature of 18 °C. Best uncorked and decanted before serving.

Tenuta Rocca (30)

In the heart of the Langhe district, in the Ornati area of Monforte d'Alba, Tenuta Rocca is an estate with 37.06 acres (15 hectares) of vineyards. From 1986, it was managed by brothers Bruno and Gianni, who constantly modernized the vineyards over the years, and since 2010 Gianni and his son Andrea, an enotechnician, have taken over management of the estate. The large farmhouse dating back to the eighteenth century has been restored to its original splendour following meticulous architectural refurbishment. The creation of the modern cellar with underground rooms for optimal development and ageing of wines confirms the commitment, passion and determination with which the wine production that symbolizes the Alba area of the Langhe district is conceived.

Barolo Docg Bussia

Barolo DOCG Bussia is made with estate-grown grapes from the Nebbiolo vineyards shown on map page no. 7, 272. Southwest exposure.

Vinification: During the ripening of the grapes we thin out the bunches to optimize the yield and then hand-select them before they undergo soft crushing. Selected yeasts that respect the fragrant aroma of the product are added to the must. At the end of temperature-controlled fermentation, the wine is drawn off and racked. The wine spends at least two years in oak barrels. It is then prepared for bottling and bottle refinement of up to a year before being released on the market, according to regulations, no sooner than the fourth year after the harvest.

Visual aspects: Brilliant, garnet red with good intensity, with slight garnet/orange highlights.
Olfactory aspects: Ethereal, intense and persistent, with scents of violet, cherry, spices (cinnamon, cloves), vanilla and leather.
Taste aspects: Dry, warm, with generous body, robust and suitably tannic but not too austere, smooth, balanced, with an intense and persistent finish characterised by sensations reminiscent of the bouquet

Aging/Method of storage: At least 18/20 years if stored correctly, away from light, at a cool and constant temperature, in the same horizontal position as when it arrives in its case until the moment of consumption.

Awards

2014 IWC Commended 2010 vintage
Wine Enthusiast – 2010 vintage   87 points

Cavallotto Tenuta Bricco Boschis (31_1)

CAVALLOTTO
Tenuta Bricco Boschis

BAROLO riserva BRICCO BOSCHIS
vigna SAN GIUSEPPE

VIGNETO: vigna SAN GIUSEPPE
cru BRICCO BOSCHIS
(nebbiolo da Barolo 100%)

Map.3 Part. n° 131, 134, 234, 276, 244, 258.
ha: 3.78
Giacitura: collina
Esposizione: Sud-Ovest
n° di viti: 18.889

Prod. Selezionata: 13333 bts+
2000 Magnum e 750 Doppi Magnum

OPERAZIONI IN VIGNETO:
Potatura a Gujot con capo a frutto di 8-10 gemme.
Inerbimento spontaneo totale. Interventi di solfato e ossido di rame (contro peronospora) e zolfo di cava in polvere (contro oidio). Nessun trattamento con insetticidi, acaricidi o altri prodotti chimici di sintesi (tranne n°1 trattamento obbligatorio come da D.M.31.05.2000 "Misure per la lotta obbligatoria contro la Flavescenza dorata"). N° 2 diradamenti grappoli: all'invaiatura (seconda-terza decade di Agosto) e in pre-vendemmia.
Vendemmia (manuale): dal 2 al 28 di Ottobre.

OPERAZIONI DI CANTINA:
diraspatura delle uve, fermentazione a temperatura controllata a 29° C massimo, macerazione a cappello semisommerso con follature in vasche orizzontali per 20-35 giorni secondo l'annata. Invecchiamento 4-5 anni secondo l'annata, in botti di rovere di Slavonia da 20-30-50-80-100 HL e almeno 6 mesi di affinamento in bottiglia coricata

ABBINAMENTI GASTRONOMICI:
Antipasti di prosciutto, insaccati, carne cruda all'albese e vitello tonnato, con secondi di carne rossa in generale, crostacei crudi e cotti, tonno, pesci e carni bianche grigliate, con formaggi a grana dura o stagionati, dolci al cioccolato o da solo da meditazione.

Cavallotto Tenuta Bricco Boschis (31_2)

CAVALLOTTO
Tenuta Bricco Boschis

BAROLO BRICCO BOSCHIS

VIGNETO:cru BRICCO BOSCHIS
(nebbiolo da Barolo 100%)

vigna Colle Sudovest: Map.3 Part. n° 187
vigna Punta Marcello: Map.3 Part. n°122,
123, 130p, 131, 216p, 217, 218p, 243
vigna San Giuseppe (parte): Map.3 Part.
n° 197, 198, 204p, 206p, 254, 255p, 256.

ha: 5.04.52,
n° di viti: 25.222
Giacitura: collina

Produzione media: 29000 bottiglie.

OPERAZIONI IN VIGNETO:

Potatura a Gujot con capo a frutto di 8-10 gemme.
Inerbimento spontaneo totale. Interventi di solfato e ossido di rame (contro
peronospora) e zolfo di cava in polvere (contro oidio). Nessun trattamento
con insetticidi, acaricidi o altri prodotti chimici di sintesi. N° 2 diradamenti
grappoli: 1° all'invaiatura (seconda-terza decade di Agosto) e il 2° prima
della vendemmia.
Vendemmia (manuale): dall'2 al 28 di Ottobre

OPERAZIONI DI CANTINA:
diraspatura delle uve di Nebbiolo da Barolo, fermentazione a temperatura
controllata a 29° C massimo, macerazione a cappello semisommerso con
follature in vasche orizzontali per 20-35 giorni secondo l'annata.
Invecchiamento di 15-18 mesi, in botti di rovere di Slavonia da 10-20-30-
50-80-100 HL e almeno 6 mesi di affinamento in bottiglia coricata.

ABBINAMENTI GASTRONOMICI:
Trova la migliore collocazione a tavola, con antipasti di prosciutto, insaccati,
carne cruda all'albese e vitello tonnato, con secondi di carne rossa in
generale, pesce e carni bianche grigliate, con formaggi a grana dura o
stagionati, dolci al cioccolato o da solo da meditazione.

Gemma (32)

Gemma Barolo Colarej 2011
Characteristics

Classification D.O.C.G.

Grape Varieties 100% Nebbiolo

HarvestIn the second ten days of October

Vinification Soft pressing and vinification in stainless steel tanks for 18 days at 26 28° C. Malolactic fermentation in stainless steel tanks.

Maturation In Slavonian oak barrels for 18 months followed by a further aging in bottle.

Tasting notes

Colour Intense ruby red with a touch of orange

Bouquet The bouquet is characterized by evident spicy notes, as well as an aroma of ripe red fruits and vanilla

Taste The taste is full, elegant, with a well balanced acidity and a good presence of tannins. This wine has a strong personality and a high potential of evolution and longevity

Pairing Red meat, game and mature cheese

Vietto (33)

Barolo "Panerole"

Productive procedure
Manual grape harvest in crates, crushing, de-stemming and fermentation "a cappello emerso" with daily pump overs, prolonged maceration, besides fermentation for 30 – 35 days, after the racking off, it is decanted various times to prepare the wine for the maleolactic fermentation and within December the ageing process in wood begins and, in this case, lasts at least three years. After the bottling, that takes place in summer, the wine evolves in the bottle, for another 18 months before entering the market.

Tasting
Vivid bright ruby red colour with slight orangey reflexes.
The nose reveals a clear ample scent with notes of violet, liquorice, cinnamon and sweet tobacco. On the palate it is full, harmonic, warm, with a pleasant and spicy finish.

Comment
An important wine, with great character and longevity. It accompanies meat based second courses or well-aged and mountain pastured cheeses.

Raineri Gianmatteo (34)

Azienda Agricola Raineri Gianmatteo
- Vineyards in Monforte d'Alba -
Loc. Panerole 24
12060 - NOVELLO

Barolo DOCG "Monserra" 2010

Kind of wine:	red
Grape variety:	nebbiolo 100%
Appellation:	Barolo DOCG
Vineyard in:	Perno of Monforte and Serralunga d'Alba
Single Vineyard:	Santo Stefano di Perno (Monforte); San Bernardo (Serralunga)
Average production:	2500 bottles; 210 cases
Grape yield per hectare:	5 tons
Exposure and altitude:	south-west, 350 mt a.s.l.
Type of soil:	sand and marls of clay and limestone
Age of vineyard and growing system:	40 year old, guyot
Harvest time:	mid-October

Farming
Grass kept short, copper and sulfur.
At least two green harvests between July and August, followed by manual selection at picking time. Grapes harvested in 20kg racks.

Vinification
Clusters are crushed and destemmed in the cellar. Fermentation and maceration last about 21 days at controlled temperature (30°-31°C; 86°-88°F) in vertical stainless steel tank.

Ageing
About two months of malolactic fermentation followed by 24 months ageing in French oak (50% new). No filtration, no fining.

Sensory characteristics
Colour: garnet-red
Nose: violet, dried rose, balsamic, mushrooms, mature fruits, tobacco
Mouth: dry, very deep and lasting, big tannins, full body, well balanced
Serving temperature: 16°-17°C; 61°-63°F
Ageing potential: 5-20 years
Pairing: pasta with rich sauce, steak, braised meat, game, aged cheese, dark chocolate

索引

Cadia (35)

AZIENDA AGRICOLA CADIA DI GIACHINO BRUNO
VIA RODDI VERDUNO, 62 - 12060 RODDI D'ALBA (CN) ITALIA
Tel. 0173 / 615398 - P. IVA 02313540045

The Process of Manufacturing

BAROLO
DENOMINAZIONE DI ORIGINE CONTROLLATA E GARANTITA
2010

Vineyards: Single Vineyards of 4400 square metres situated in the village commune area of Roddi in the subzone known as "Monvigliero" placed at a height of 380 metres above sea level with a density of the vines of 5300 per hectare.

Soil: Loose and calcareous of medium structure.

Grapes: Nebbiolo 100%

Yield: 7.000 Kg of grapes per Ha = 45.5 Hl of wine per Ha.

Harvest: picking by hand with a first careful selection of the grapes in the vineyard
The harvest was carried out on October 4th 2010.

Vinification: the grapes, after the first selection in the field, were soft pressed and destalked and placed in temperature controlled stainless steel tanks to ferment for a period of 15 days at an average temperature of 25-28° C., Once the fermentation is completed the must-wine is gentle pressed with air pressers to obtain the first press Once the fermentation is completed the wine is divided from the skins that are soft pressed with air pressers so to o obtain the first press.

Ageing:2 years in 225 litres French "Allier" " medium toasted fine grain oak barrique ; after the wine is fined in the bottle for 1 year after bottling before being put on the market but without no doubt the wine will improve for several years in the bottle.

Bottling: March 05th 2014
Alcohol: 14,67% by vol.
Acidity: 5,71 gr/lt
Extract: 27,60gr/lt
Sugar: 0,3 gr/lt

Paolo Manzone (36_1)

Meriame

ALCOHOL DEGREE: 14% vol

ACIDITY: 5,4 gr/Lt in tartaric acid

GRAPE VARIETY: 100% Nebbiolo.

PRODUCTION AREA: This is a traditional cru area of Barolo, situated in the south facing slope of a hill in the village of Serralunga d'Alba. The places and hillsides for the cultivation of Barolo "the wine of kings and the king of wines" have been selected down the centuries.

AGE OF VINEYARD: 60 years old vines in Serralunga d'Alba.

PRODUCT YIELD: 60 Q.li per ha. as to the grapes and 40 hl .per ha. as to the wine.

GROWING SYSTEM: Typical espalier hill cultivation, Goujot pruning withj ust one fruity vine stalk of 8 buds corresponding to 6 productive ones.

DENSITY: Around 4. 500 vines per ha.

SOIL: The soil of cultivation is hilly type, between 250 and 300 meter high above sea level; south-south-east located with a mixed clayish and calcareous structure. This particular position makes a rapid maturation possible especially if combined with the thinning out of the bunches of grapes. The vinqvards are situated in one of the best areas specifically adapt to this particular cru. Infact this clay limestone soil, which is poor but perfectly fit to vines can befound on an amphitheatre shaped area which attracts heat and concentrates it by maintaining it. The perfect exposition allows a total exploitation of sunlight. Then, this particular "boul" protects the vines from the wind and prevents from any sudden temperature falls.

VINIFICATION: After picking of the stalks, the free run juices are stocked in vertical stainless steel vats. The fermentation takes place at controlled temperature of 27/30 C. The maceration process is quite long 14 days average. The wine was settled in November in the tanks, before being poured of and aged in french tonneaux. No wine more than Barolo gets better with time.

COMBINING FOODS AND FLAVOURS: Full, austere intense with signs of violet and dried canine rose; red intense pomegranate with typical orange reflexes. No wine more than Barolo gets better with time. It should be served with the great kitchen dishes, but it admirably closes also a meal as meditation wine.

Paolo Manzone (36_2)

Serralunga

ALCOHOL DEGREE: 13,5% vol

ACIDITY: 5,6 gr/Lt in tartaric acid

GRAPE VARIETY: 100% Nebbiolo.

PRODUCTION AREA: This is a traditional cru area of Barolo, situated in the south facing slope ofa hill in the village of Serralunga d'Alba. The places and hillsides for the cultivation of Barolo "the wine of kings and the king of wines" have been selected down the centuries.

AGE OF VINEYARD: They are relatively young vineyards, 15years, set at a height of about 400 mt. In Serralunga d'Alba.

PRODUCT YIELD: 60 Q.li per ha. as to the grapes and 40 hl. per ha. as to the wine.

GROWING SYSTEM: Typical espalier hill cultivation, Goujot pruning with just one fruity vine stalk of 8 buds corresponding to 6 productive ones.

DENSITY: Around 4.500 vines per ha.

SOIL: The soil of cultivation is hilly type, between 250 and 300 meter high above sea level; south-south-east located with a mixed clayish and calcareous structure. This particular position makes a rapid maturation possible especially combined with the thinning out of the bunches of grapes. The vineyards are situated in one of the best areas specifically adapt to this particular cru. Infact this clay-limestone soil, which is poor but perfectly fit to vines can be found on an amphitheatre shaped area which attracts heat and concentrates it by maintaining it The perfect exposition allows a total exploitation of sunlight. Then, this particular "boul" protects the vines from the wind and prevents from any sudden temperaturefalls.

VINIFICATION: After picking off the stalks, the free run juices are stocked in vertical stainless steel vats. The fermentation takes place at controlled temperature of 27/30 C. The maceration process is quite long 12 days average. The wine was settled in November in the tanks, before being poured of and aged in french tonneaax. No wine more than Barolo gets better with time.

COMBINING FOODS AND FLAVOURS: It should be served with the great kitchen dishes, but it admirably closes also a meal as meditation wine.

Gianni Ramello (37)

After studyng at the school desk and years spent as a construction site worker, I finally managed to realise my dream in 2000.

The new millennium saw me working in the vineyards that were once my grandfather's and then my father's and hence being born as a producer.

Bringing together tradition and technique I aim to produce wines which reflect our land, the Langhe, in the best possible way. Their intense aromas remind us of the springtime whilst their structure reminds us of the strenuous vineyard work.

Knowing that in this type of work nothing should be taking for granted and even working in the same way each vineyard and each barrel are different, the vintage then adds its own touch, trasforming our work into a continuous challenge to produce wine which macth our high expectations each year.

My business isn't big, just three hectares of vineyard, where I grow Dolcetto in Pisote, Nebbiolo da Barolo in Rocchettevino and Barbera in Monticello, on the sandy soil left of the River Tanaro.

BAROLO

This is produced with the Nebbiolo grape from the vines in the area of **Rocchettevino**, part of town of La Morra.

The harvest starts during the second half of October, after the pressing, the fermentation process begins which stretches over fifteen to twenty days. Once that is finished the new wine is decanted for some days and then immediately poured into wooden casks where it remains to age for at least two years.

In the summer of the third year the wine is filtered and bottled, and after at least six months refinement in the bottle the Barolo is ready to be consumed.

The wine has a ruby red colour with orange reflections. The young Barolo (2-4 years old) has a spacius floral bouquet (wild rose and vanilla) which develops into a more spicy bouquet (leather, liquorice) in an older Barolo.

Wines produced

Barolo ROCCHETTEVINO	7.000	bottles
Dolcetto d'Alba PISOTE	1.000	bottles
Dolcetto d'Alba	5.000	bottles
Barbera d'Alba	1.000	bottles

cultivated vineyards

Nebbiolo da Barolo Rocchettevino vineyard	Ha	1,10
Dolcetto d'Alba Pisote vineyard	Ha	1,30
Barbera d'Alba Mentin vineyard	Ha	0,20

索引

Marchesi di Barolo (38_1)

GRAPE VARIETY:
Nebbiolo 100%

AREA OF PRODUCTION:
Cannubi is a long hill with a gradual slope lying in the heart of the Barolo area.
Here the Helvetian and the Tortonian soils blend together and originate gray-blue marls rich in magnesium and manganese carbonate that, on the surface, thanks to the air and the weather, turn into grey-white marls: they are made of a combination of sand, lime, calcareous and clay (marls of Sant'Agata fossils). Surrounded by higher hills, the Cannubi hill is protected from storms and high winds and it benefits from an unusual and unique microclimate. The characteristics of the soil and the extraordinary microclimate give Cannubi an exceptional completeness and balance, harmony between structure and aromas and remarkably elegant tannins, making this Barolo not just a pleasure to drink but also a wine with a long lifespan. The vineyards, East / South-East exposed, are mapped on the sheet n.7 of the municipality of Barolo, lots 44, 170, 219, 231, 235, 242.

TRAINING SYSTEM:
Low Guyot trellised system is used with a density of 4,000 vines per hectare.

VINIFICATION:
The grapes are collected and quickly taken to the cellar where they are destalked and softly pressed to extract from the peel and the outer area of the grape only the most noble and aromatic notes. Fermentation, at controlled temperature, takes place in thermo-conditioned tanks. Maceration lasts for 10 days. During this time, the wine is regularly recycled from the bottom to the top of the tank in order to take all the elements present on the skins and to extract the color slowly and softly. Once the fermentation is finished the natural sugars of the grape are totally converted into alcohol. The wine is then racked into cement tanks that are lined with fiberglass and isolated by cork. Here the wine keeps the post-fermentation temperature of 22° C for a long time. In this way malolactic fermentation starts spontaneously, ending after two months.

AGEING:
The wine ages for two years, a part in Slavonian oak barrels (30 or 35 hectoliters / 789-947 U.S. gallons) and the other part in small French oak barrels (225 liters) that are moderately toasted. The two parts are then blended in traditional big oak barrels and the wine completes its fining in the bottle for 12 months, before going into the market. Barolo Cannubi reaches maturity after 6 years from harvest and the ripening plateau is between 6 and 25 years.

ORGANOLEPTIC CHARACTERISTICS:
Garnet-red in color with ruby reflections. Intense perfume with clean scent of roses, vanilla, licorice, spices and toasted oak. Gentle note of absinth. The flavor is full and elegant, good-bodied and austere with remarkably elegant tannins. The spicy note and the hints of wood blend perfectly.

PAIRINGS:
Barolo Cannubi is well-matched with traditional Langhe egg pasta, Tajarin and Ravioli del Plin, with roasts, stews, braised meats and game. Ideal companion also for goat milk cheeses and mildly-aged cheeses.

SERVING TEMPERATURE:
18° C (64° F)

ALCOHOL BY VOLUME:
14.5% Vol

Marchesi di Barolo (38_2)

VARIETY
Nebbiolo 100%

PRODUCTION ZONE:
The southeast-exposed Sarmassa vineyard is located on a hill with good slope. Despite the fact that the surrounding area is of Tortonian origin because of a significant soil erosion due to the steep slope of the hill, the soil has the typical characteristics of grounds of Elvetian origin: it is composed mainly of clay and limestone and has a very substantial percentage of stones. The high presence of stones, combined with clay, limits the growth of the Nebbiolo grape and allow the vines to react quickly to climatic variations, enabling clusters to achieve a perfect ripening. Mapped to the sheet number 9 of the commune of Barolo, particles 104, 149, 157, 303.

CULTIVATION OF THE VINEYARD:
Low Guyot trellised system with a density of 4,000 vines per hectare.

VINIFICATION:
The grapes are collected and quickly taken to the cellar where they are destalked and softly pressed to extract from the peel and the outer area of the grape only the most noble and aromatic notes. Fermentation, at controlled temperature, takes place in thermo-conditioned tanks. Maceration lasts for 10 days. During this time the wine is regularly recycled from the bottom to the top of the tank in order to extract all the elements present on the skins as well as color slowly and softly. Once fermentation is finished the natural sugars of the grape are totally converted into alcohol. The wine is then racked into cement vats lined with fiber glass and isolated by cork in order to keep the post-fermentation temperature of 22° degrees C (72° F) for a long time. In this way, malolactic fermentation starts spontaneously, ending after two months.

AGING:
The wine ages for two years: a part in Slavonian oak barrels of 30 or 35 hectoliters (789-947 U.S. gallons) and the other part in small French medium-toasted oak barrels (225 liters). The two parts are then blended in traditional big oak barrels and the wine completes its fining in the bottle before going into the market. Barolo Sarmassa reaches maturity after 8 years from harvest and the ripening plateau is between 8 and 30 years. Therefore, this wine is well-structured, colorful, tannic and long-lived.

SENSORY CHARACTERISTICS:
Deep garnet red. Intense aroma with clear scents of wild rose, vanilla, licorice and spices. Delicate perfumes of pine resin and tobacco. The taste is full and elegant, full-bodied, with tannins in evidence. The pleasant spicy and woody notes blend perfectly.

FOOD MATCHES:
Barolo Sarmassa is well-matched with traditional Langhe egg pasta, tajarin and ravioli del plin, with roasts, stews, braised meats and game. Ideal companion also for goat milk cheeses and mildly-aged cheeses.

SERVING TEMPERATURE:
18°-20° degrees C (64-68° F)

ALCOHOL DEGREES:
14% Vol

Marchesi di Barolo (38_3)

GRAPE VARIETY:
Nebbiolo 100%

AREA OF PRODUCTION:
Coste di Rose is a prestigious hill with full eastern exposure located in the municipality of Barolo. Coste di Rose is characterized by a very steep slope (more than 40%) which allows the perfect maturation of the Nebbiolo clusters, extremely demanding in terms of light and heat. Even if the hill is from the Helvetian geological age, curiously the soil quality is moderately calcareous (this originated from the marine deposits uplift, which are still found through regular agricultural maintenance) and very rich in quartzite sand, fine limestone and just a small amount of clay. These peculiar characteristics confer to the Nebbiolo vines an intense and fine aroma that brings to mind hints of wild mint.

TRAINING SYSTEM:
Low Guyot trellised system is used.

VINIFICATION:
The grapes are collected and quickly taken to the cellar where they are destalked and softly pressed to extract from the peel and the outer area of the grape only the most noble and aromatic notes. Fermentation, at controlled temperature, takes place in thermo-conditioned tanks. Maceration lasts for 10 days. During this time the fermenting wine is regularly recycled from the bottom to the top of the tank in order to take all the elements present on the skins and to extract the color slowly and softly. Once fermentation is finished the natural sugars of the grape are totally converted into alcohol. The wine is then racked into cement vats lined with fiberglass and isolated by cork. Here the wine keeps the post-fermentation temperature of 22° C for a long time. In this way, malolactic fermentation starts spontaneously, ending after two months.

AGEING:
The wine is initially aged for 2 years: a part in Slavonian oak barrels (30 or 36 hectoliters / 789 or 947 U.S. gallons), and the other part in small, slightly toasted, French oak barrels (225 liters / 59 U.S. gallons). The vineyard finds again its unity by assembling the wine in the traditional big oak barrels and ends its fining in the bottle for 12 months, before going into the market. Barolo Coste di Rose reaches its maturity after 4 years from harvest and the ripening plateau is between 4 and 20 years.

ORGANOLEPTIC CHARACTERISTICS:
Ruby-red color tending to garnet. Intense perfume with clean scents of roses, liquorice, spices and aromatic herbs, in particular wild mint. Full, elegant, and full-bodied flavor. The soft color and the structure confirm it as an immediately pleasant, balanced and harmonious Barolo.

PAIRINGS:
This wine is well-matched with typical appetizers from Langhe: raw meat, the vegetable flan, eggs with truffles. Excellent with tajarin (traditional Langhe egg pasta), roasts, cheeses of sheep and goat milk and mildly seasoned cheeses.

SERVING TEMPERATURE:
18° C (64° F)

ALCOHOL BY VOLUME:
14.5% Vol

anbiagio (39_1)

Barolo *Sorano* docg

Grape-variety: Nebbiolo

Growing location: first produced in the 2005 vintage, Barolo Sorano is grown on two parcels of land located in the higher part of Sorano, an outlying district of the village of Serralunga d'Alba. The vineyards are between 15 and 25 years of age, and face south-east and south-west on soil composed of clayey and calcareous marl.

Harvest: in October. Before picking attentive thinning of the bunches takes place throughout the month of August. This is followed by a painstaking selection of only the best grapes during the picking, and again in the winery prior to crushing. The yield never exceeds 5.5 tons per hectare.

Vinification: in order to bring out the best in the richness of the grapes grown on this important site, it was decided to employ a strict style of vinification that would be capable of giving this wine the unique sense of "terroir" this cru is capable of expressing. This led to the rediscovery of old methods of vinification used in days gone by: the grapes are crushed in large vats made of Slavonian oak, and the juice is left to macerate on the skins for 2 days. This is followed by a whole week of delicate breaking up of the cap to make the mass homogeneous. The pomace is then covered with the wine, which is left to ferment slowly for more than 2 months, allowing the fermentative-enzymatic processes to extract the polyphenols and provide a rich aromatic structure.

Maturing: after the vinification, the wine matures in the same 10,000-litre vats for around 30 months, followed by a further period of 12 months in the bottle. During this ageing there is a clear-cut evolution of the tertiary aromas.

Colour: very intense, bright garnet red, tending to more mature shades on ageing.

Nose: complex, with floral notes and black-berried fruit, including ripe blackberry. Spicy aromas - especially tobacco - follow, along with underbrush and tar: lingering tertiary aromas tending to the ethereal.

Taste: the qualities provided by the soil in Serralunga d'Alba show through quite clearly on the palate, with appealing full flavour, a well-balanced structure, and balsamic notes that merge beautifully with the sweetness of the tannins. Nice hints of liquorice, tobacco and tar on the finish.

Pairings: at its very best with roasts, braised meats, game, and medium and mature cheeses.

Sanbiagio (39_2)

Barolo Bricco San Biagio docg

Grape-variety: Nebbiolo

Growing location: Bricco San Biagio is produced from a historic Nebbiolo vineyard lying at the heart of one of the finest Barolo crus, the San Biagio hill in the village of La Morra. The particular hollow-shaped conformation of the slope enables the warmth captured during the day to be retained over the night. The vineyard is named after the round hillnet - "Bricco" in Italian - that stands at the top of the vineyard, dominating the landscape. Facing east, south-east, its calcareous, clayey soil is of Miocene origin.

Harvest: the advanced age of the vineyard and the particular vegetative vigour of the vines contribute to producing grapes of the highest quality. Heavy thinning of the bunches over the summer brings the harvest forward to mid-October, when the hand-picked grapes undergo careful selection both in the vineyard and again on crushing. In keeping with the estate's ambitious philosophy, the yield is never higher than 5 tons/Ha, well below the maximum limit of 8 tons/Ha set by production regulations.

Vinification: the maceration on the skins lasts around 20 days in small new oak vats, avoiding overheating of the pomace and any unpleasant 'cooked' overtones that could result. Frequent, gentle breaking up of the cap by hand during this period allows for good extraction of the finer polyphenols that provide softness, and stimulates the enzymatic and fermentation processes that can give this wine the unique sense of "terroir" this cru is capable of expressing. Following the alcoholic fermentation, the vats are emptied, washed and closed. At this stage, the wine is still turbid and full of yeasts, so it is put back into the same casks for the malolactic fermentation, and is left there on the lees until March. This sets in motion highly complex enzymatic-fermentative processes that guarantee the development of a wealth of aromas and give the wine a fatter feel.

Maturing: in Slavonian oak casks for no less than 30 months. Ageing in the bottle follows for 1 year, during which the tannins soften, and the floral and fruity notes develop.

Colour: deep garnet red, tending to more mature shades on ageing.

Nose: expansive, featuring berries, dog roses, lime, jam, cooked plums. Appealing spicy and balsamic overtones stand out in the more mature wine, especially star anise, liquorice, mushrooms and dried figs, as well as hints of tobacco and leather.

Taste: strikingly complex, with spicy and balsamic notes and sweet tannins clearly showing through. Appealing overtones of liquorice, tobacco and tar on the finish.

anbiagio (39_3)

Barolo *"Rocchettevino"* docg

Grape-variety: Nebbiolo

Growing location: the Rocchettevino cru was planted in 1997, on the slopes of an east-facing hillside in the village of La Morra at an average altitude of 300 metres. "Ronchetovinum" was cited in the land registry as long ago as 1477 as one of the very finest wine growing sites in the village. The soil features clayey-calcareous marl, which contributes to bringing out this wine's aromatic finesse.

Harvest: the beginning of October. This cru, which is distinguished by particular vegetative vigour, produces grapes of the highest quality. These prized characteristics are further guaranteed by thinning of the bunches by hand between the end of July and the beginning of September, and by strict selection on crushing. The yield is always between 5 and 6.5 tons per hectare.

Vinification: the maceration on the skins lasts 30-35 days, taking place in heat-conditioned steel tanks to avoid the risk of the pomace overheating and causing unpleasant 'cooked' overtones. Frequent, gentle breaking up the cap guarantees remixing of the mass, stimulating the enzymatic and fermentation processes, and encouraging the development of the aromatic fraction, in particular fruity aromas. These operations also cater for achieving good extraction of the finer polyphenolic substances that give the wine sweetness and softness.

Maturing: in casks of French oak holding 700 litres for around 27 months, followed by a further 15 months in the bottle.

Colour: bright, deep garnet red.

Nose: elegant, complex. Balsamic and floral aromas. Stands out for the freshness of its bouquet and the ethereal overtones that increase with ageing.

Taste: the qualities showing on the nose are confirmed. Soft tannins, expansive sensations of sweetness, aromatic notes.

Recommended pairings: at its very best with roast and braised meats, game, and medium and mature cheeses.

索引

Fratelli Giacosa (40)

Barolo

Giacosa Fratelli
neive - Italia

Bussia
denomination of controlled and guaranteed origin

Grown in vineyards in the subarea of Bussia in Monforte d'Alba. The balance between the clayey and calcareous elements in the soil, the high presence of magnesium and microelements, the temperature ranges of the altitude and location of the valley mouth favour the essentials, in these grapes, of particular fruity notes.

Non-invasive winemaking, lengthy maceration, and ageing in casks of 30 and 60 hectolitres (60 and 120 kegs) produce a velvety wine with smooth tanicity.

Scarzello Giorgio (41)

SCARZELLO GIORGIO & FIGLI

Barolo DOCG Sarmassa Vigna Merenda 2008

Produced with Nebbiolo grapes from Vigna Merenda, in the hearth of Sarmassa area

Altitude 300/350 m.s.l.

Exposure South – South East

Total Surface 2 hectares

Soil Calcareous and clayey

Growing system espalier with Guyot pruning

Density 5000 plants per hectare

Yield 6,5/7,0 tons per hectare

Average age of the vines 10/15 years

Harvest manual in the first half of October.

First Vintage produced 1978

Bottles produced in 2008 5000

Alcohol 14,5% vol.

Vinification and ageing maceration of about 50 days, refinement in 25Hl barrels of Slavonian oak for 30 months and two years ageing in bottle.

Notes

During harvest the four parcels of Vigna Merenda vineyard are vinificated separetly. After one year of refinement in wood, the four wines are blended and we decide the best blending right for obtaining an excellent Barolo, and for that produced only in the best vintage.
Vigna Merenda is a traditional Barolo, a wine good for long ageing, which expresses at its best the characteristics of Barolo terroir. Thanks to its power and tannins balanced by an equilibrated harmony, it's an extremely elegant wine,
It's more advisable to decant the wine for better exalting all the nuances of its aromas. The best match is with tasty meat main courses

Josetta Saffirio (43)

BAROLO
Denominazione di Origine Controllata e Garantita

GRAPE VARIETY: Nebbiolo 100% (Michet).

PRODUCTION AREA: Comune of Monforte d'Alba – subappellation Castelletto. Soil type here is called 'Elveziano' and is best described as marine deposits that have been pushed upwards to today's altitudes, showing traces of fossils from the Miocene, lime, clay and sand. Very rich in minerals and live lime.

VINEYARD TRELLACING: Low Guyot system on hills with good inclines (40-50%) with south-est exposure.

YIELD PER HECTARE: 45 hectolitres.

VINIFICATION: Soft pressing of the hand-harvested grapes with destalking, followed by fermentation in temperature controlled tanks at 30/32 degrees C (86/88 F).

The skins are macerated for 10 days and the wine is racked when fermentation has been completed.

MATURATION: After malolactic fermentation in december, the wine is poured in barrels. It is decanted once a year and after 24 months is poured in cement tanks and bottled in the following spring.

SENSORY CHARACTERISTICS: Garnet-red in color with ruby reflections. An intense odour with clean scents of roses, vanilla, roasted hazelnuts, licorice, tobacco and cinnamon. Fruity clean scents with strawberry and mulberry flavour.

A full, elegant and austere flavour with a good body and recurring olfactory sensations. The perfectly blended hints of fruits and wood are extremely appealing.

FOOD MATCHES: With its big structure, this wine is particulary adapted to main courses of red meats, braised dishes and game in general. An ideal accompainement for cheeses.

SERVING TEMPERATURE: 18 degrees C.

CELLERAGE: It is recommended that the bottle be laid down in a cool, airy place protected from all sources of light and far from noises and vibrations.

Ciabot Berton (**44_1**)

BAROLO Roggeri D.O.C.G.

variety: Nebbiolo 100%
Vineyards: located in La Morra, in the Roggeri cru'
Exposure: hilly, with south-east solar exposure, at 300 m above sea level
Terrain: clayey-calcareous
Yield per hectar: 40 hectoliters (hl)
Average age of vineyards: 40 years
Density/hectar: 4000 vines
Annual production: 8.000 bottles of 0,75 liters

Wine making: traditional method with maceration of the must on the skins for at least 18-20 days at 30°C in fiberglass lined concrete vats.
Ageing: the wine is aged first in French oak casks, then in Slavonia oak casks and bottled thereafter.
Alcohol content: 14,0 -14,5% by vol.
Acidity: 5,4 - 5,6 grams/liter.
Longevity: 12-15 years

Tasting notes: intense red garnet color, with ruby reflexions, decidedly fruity aroma, with scents of prunes, dried flowers and spices. Full bodied to the taste, with strong tannic but silky structure. Finish persistently long.

Gastronomic combinations: it is excellent with game, braised beef, seasoned cheeses and all dishes served with white truffles.

Ciabot Berton (44_2)

CIABOT BERTON

La Morra
Piemonte
Italia

BAROLO Rocchettevino

Variety: Nebbiolo 100%

Vineyards: located in LaMorra, in the Rocchettevino cru

Exposure: hilly, with solar exposure to the south-east, at 400 m. above sea level

Terrain: clayey- calcareous

Yield per hectar: 45 hl.

Average age of vineyards: 15 years

Density/hectar: 4000 vines

Annual production: 3.000 0,75 l. bottles

Wine making: maceration at 30-32°C for 18-20 days in fiberglass lined concrete vats

Ageing: 24 months in 50 hl. Slavonia oak casks

Alcohol content: 14,0 -14,5 by vol.

Acidity: 5,4 - 5,6 grams/liter

Longevity: 12-15 years

Tasting notes: bright ruby-red hue with light garnet reflections. Fruity notes perfectly blending in a flowery, spicy aroma make the wine extremely fine. Complex and clean in mouth, the wine shows a firm structure supported by a lively acidity that are well-integrated with the typical tannic sensations of Nebbiolo grapes. Long-lasting finish.

Gastronomic combinations: Excellent with braised or roasted meat, game dishes and tasty cheese.

Bussia Soprana (45)

BUSSIA SOPRANA

BAROLO DOCG VIGNA COLONNELLO 2009

Vitigno:	nebbiolo 100%
Zona di provenienza:	Bussia Soprana,parte unica di vigneto di circa 0,5 ettari
Terreno:	misto tufo e sabbia
Densita' di impianto:	4000 ceppi per ettaro
Sistema di allevamento:	guyot
Eta' media delle viti:	60 anni
Resa per ettaro:	40/50 quintali
Gradazione alcolica:	14,°
Epoca di vendemmia:	ottobre
Materiale di fermentazione:	vasche inox
Temperatura di fermentazione:	28°/32°
Macerazione:	15 giorni circa
Lieviti impiegati:	selezionati
Invecchiamento:	botte da 30 quintali per circa 24 mesi
Bottiglie prodotte:	3.200 circa

Luigi Baudana_G.D.Vajra (46_1)

Barolo DOCG Bricco delle Viole   **G.D.VAJRA**

GRAPE VARIETY

100% Nebbiolo

DESCRIPTION

A smooth interwoven silkiness is what best characterizes Bricco delle Viole. Coming from its namesake vineyard, it is located on the highest part of the Comune of Barolo. From its uniquely high position you can see, and almost touch, the stunning Alpine mountains: a *Barolo di Barolo* where the characteristic notes of Nebbiolo settle into its most delicate and elegant version.

VINEYARDS

Bricco Delle Viole is a hill which emerged about 11 million years ago, during the geological Tortonian Age. This has resulted in a clay-calcium rich soil with a high concentration of magnesium and manganese, an altitude of 400/480 meters, a microclimate which is neither too hot in summer nor too cold in winter. Only the oldest vines are vinified for this selection, while the younger ones are champing at the bit, waiting anxiously to grow up and become "of age".

FARMING

Vajra has always believed in sustainable farming and keeps the turf natural and green in the vineyard. No irrigation is used while thinning is decided according to the vintage.

HARVEST AND WINEMAKING

These vineyards are the last to be harvested. The wait makes our hearts palpitate because the weather, in autumn, can instantly change. But it is a waiting period that is rewarded with high aromatic complexity and extreme finesse. Some years it is even necessary to patiently pick grapes in different stages, in order to respect these old vines and select the ripest grapes possible.

30 to sometimes 40 days of fermentation are needed to let the wealth of tannins pass from the skins, seeds and in some years, even the stems. A long aging period (approx. 42-48 months) in big barrels (25-50 Hl) of Slavonian oak is carried out before bottling.

TASTING NOTES

Brilliant ruby red with notes of cherries, brandied cherries, dark berries, violets and dried roses which recall nuances of citrus fruits and mint. On the palate noble and silky tannins. Very balanced with a long finish of fresh, delicate spices.

Luigi Baudana_G.D.Vajra (46_2)

GRAPE VARIETY
100% Nebbiolo

DESCRIPTION
The confluence between Tortonian and Helvetian soils delivers a Barolo of lifted aromatic and vibrant energy. Freshly flowers and delicate spices all take shape in this purest version of Barolo.

VINEYARDS
Ravera is a cru located in the southwest quadrant of the Barolo zone. The vineyard runs north to south across Commune of Barolo and Novello. The soil here is a mix of clay and sand, especially rich in iron, with a good permeability and drainage. This translates into a dense wine that promises an evolution over time. The vineyard has been farmed since 2001 but produced Langhe Nebbiolo until 2008. Today it give us a Barolo with fine aromas, sapid flavour and great complexity.

FARMING
Vajra has always believed in sustainable farming and keeps the turf natural and green in the vineyard. No irrigation is used while thinning is decided according to the vintage.

HARVEST AND WINEMAKING
The manual harvest takes place in the second/third week of October.
After that 30 to sometimes 40 days of fermentation are needed to let colour and tannins passing from the skins, seeds and in some years, even the stems. A long aging period (approx. 42 months) in big barrels (25-50 Hl) of Slavonian oak is carried out before bottling.

TASTING NOTES
Brilliant ruby red with notes of dark fruits, deep cherries, new leather, dark spices. On the palate noble and long tannins. Very balanced and full bodied with ripe juicy cherry and a durable orange-like finish.

Luigi Baudana_G.D.Vajra (46_3)

LUIGI BAUDANA
SERRALUNGA D'ALBA

BAROLO DOCG BAUDANA

COMUNE DI SERRALUNGA D'ALBA

Western Slope – first sector
Bordering: Comune di Castiglione Falletto
Size of cru: medium-small
Altitude: approx. 250-350 m.
Exposures: south to south-west
Labels: 3 Barolo labels
Google Earth coordinates: 44°37'37.20''N 7°59'36.41''E

VINEYARDS ON THE WESTERN SLOPE FROM NORTH TO SOUTH:
Sorano - Baudana – Parafada – Lazzarito – Rionda – Ornato – Falletto -

Castiglione Falletto

Baudana is an historical vineyard, a *treasure island* whose existence is almost forgotten. From the Atlante delle Vigne di Langa (Slow Food, 1990): *"we can assert that the Baudana vineyard contributed to create Barolos of excellent signatures since the Thirties: this can't but uphold the consolidated opinion of quality expressed by the locals"*. Most of its fruit is today sold to large cooperatives and not estate bottled, thus producing an oblivion that can't hide its qualities.

The area lays at the border between the fossils *Marne di Sant'Agata* and the *Formazioni di Lequio,* with a compact, calcareous, white soil. The limited percentage of sands enhances water retention in the driest vintages.

Two vineyards: vineyard 'A' is located in *Surì Baudana*, the steepest slope of the cru. Vineyard 'B' is bending on a sheltered amphitheater bending at south-west. The total surface of the vineyards is 1 hectare, with a density of 4.200 plants.

Baudana produces rich Barolos of wide structure and a spur of smoothness to the finish.

Harvest is anticipated by multiple trimming on the vine, according to the vintage. Thanks to the balanced ground conditions, ripening is smooth and fruit is usually picked at mid of October.

Fermentation takes place in oak vats and stainless-steel tanks. Manual punch-down and gentle pumping-overs are the only operations over the time of maceration. Compared to Cerretta, Baudana fruit tends to speed up the beginning of fermentation, yet with a long, steady evolution. As a consequence, maceration may last as long as 30 days. A post-fermentation 'cappello sommerso' [submerged marc] will follow for another week in the greatest vintages.

The wine is wracked into oak after malolactic fermentation. Since vintage 2005, the production of this Barolo has been reduced of about 40%. Only the best grapes make it into the bottle, while fruit from younger plants is blended in the classic Barolo Serralunga d'Alba. From the same year, Luigi Baudana moved to a longer ageing process which includes Slavonian oak and an extensive ageing up to thirty-six months prior to bottling.

Luigi Baudana is the last scion of one of the oldest wine families of Serralunga d'Alba. His winery has been a secret for a bunch of fortunate wine lovers for over thirty years, producing a limited amount of Barolo and other wines. In 2003 Luigi and wife Fiorina started to look for a future for their estate. The research lasted long, until they found in Giuseppe, Francesca and Isidoro Vaira, children of Aldo and Milena of G.D. Vajra, Barolo, the enthusiasm, energies and will to preserve the past history and prolong the future of this beautiful little jewel. The three kids, in their early twenties, consider themselves '*i guardiani*' [the guardians] of this century-long winery and have started operations in 2008 vintage in the Baudana cellar. Farming is sustainable and takes advantage of organic and biodynamic practices; no chemical manipulation occurs to the wines. The average production is made of 300 cases of Barolo DOCG Cerretta and 300cs of Barolo DOCG Baudana, together with a very small bottling of Langhe made of Chardonnay (95%) and Sauvignon (5%).

Luigi Baudana_G.D.Vajra (46_4)

LUIGI BAUDANA
SERRALUNGA D'ALBA

BAROLO DOCG CERRETTA

COMUNE DI SERRALUNGA D'ALBA

Eastern Slope – first sector
Bordering: Comune di Diano d'Alba
Size of cru: large
Altitude: approx. 250-395 m.
Exposures: north-west to south-east
Labels: approx. 8 Barolo labels
Google Earth coordinates: 44°37'34.77''N 7°59'52.94''E

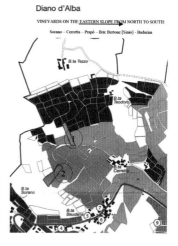

Diano d'Alba

VINEYARDS ON THE EASTERN SLOPE FROM NORTH TO SOUTH:
Sorano – Cerretta – Prapò – Bric Bertone [Simio] – Badarina

Two vineyards: vineyard '1' is located on a south-east slope; vineyard '2' is bending on a south-west slope in the once-called *Cerretta Piani* subregion. Both vineyards lay on the typical soil of the Eastern bank of Serralunga d'Alba. Here topsoil shows a very poor organic content: the consequent lack of superficial moisture forces the vine roots to dig deep in the ground. Subsoil is primarily formed by a mix of yellow sands and limestone, which refer to the Helvetian geological era. The total surface of the vineyards is of approximately 0,8 hectares, at a maximum distance of 600m from the winery. Plantings occurred in 1970 with a density of 4.000+ vines per hectare [*Luigi Baudana had an impressive vision for those years*].
The result of these conditions is an austere and very mineral expression of Barolo, with hints of herbal, cocoa powder, coffee, tar.

Fruit is cleaned and trimmed on the vine twice at least prior to the harvest, which is usually slightly anticipated in Cerretta because of the lower altitude. As an average, the crushing is completed by the second week of October.
Fermentation takes place in oak vats and stainless-steel tanks. Manual punch-down and gentle pumping-overs are the only operations over the time of maceration. The Cerretta vineyard distinguishes for a very slow beginning of fermentation, with an extended time of pre-maceration at low temperature. Indigenous yeasts are the only responsible for the long fermentation. A post-fermentation 'cappello sommerso' [submerged marc] will follow for another week in the greatest vintages.
The wine is wracked into oak after malolactic fermentation. Back until 2004 vintage the wine was aged in new and second passage French barriques for two years. Since 2005 vintage Luigi Baudana started a more gentle ageing process which includes Slavonian oak and an extensive ageing up to thirty-six months prior to bottling.

Luigi Baudana is the last scion of one of the oldest wine families of Serralunga d'Alba. His winery has been a secret for a bunch of fortunate wine lovers for over thirty years, producing a limited amount of Barolo and other wines. In 2003 Luigi and wife Fiorina started to look for a future for their estate. The research lasted long, until they found in Giuseppe, Francesca and Isidoro Vaira, children of Aldo and Milena of G.D. Vajra, Barolo, the enthusiasm, energies and will to preserve the past history and prolong the future of this beautiful little jewel. The three kids, in their early twenties, consider themselves '*i guardiani*' [the guardians] of this century-long winery and have started operations in 2008 vintage in the Baudana cellar. Farming is sustainable and takes advantage of organic and biodynamic practices; no chemical manipulation occur to the wines. The average production is made of 300 cases of Barolo DOCG Cerretta and 300cs of Barolo DOCG Baudana, together with a very small bottling of Langhe made of Chardonnay (95%) and Sauvignon (5%).

Luigi Baudana_G.D.Vajra (46_5)

GRAPE VARIETY

100% Nebbiolo

DESCRIPTION

A true *Barolo di Barolo*, Albe is a blend of grapes from three different vineyards at three different altitudes. Every morning the sun rises from one vineyard to another creating a symphony for the senses just like the wine produced from these blocks. This is a Barolo of great character and refined balance, where beautiful aromas, refined equilibrium and ripe tannins, come together in perfect harmony.

VINEYARD

With altitudes from 400 to 440 meters, different exposures and soils, the three vineyards in *Barolo, Fossati, Coste di Vergne* and *La Volta* give life to a single harmonious wine with loads of personality.

FARMING

Vajra has always believed in sustainable farming and keeps the turf natural and green in the vineyard. No irrigation is used while thinning is decided according to the vintage.

HARVEST AND WINEMAKING

The harvesting of Nebbiolo grapes in the Comune of Barolo is done during the second half of October. Grape picking and transporting the grapes in small crates are done manually to keep the grapes intact. The careful selection of grape clusters starts in the vineyard and is carried out with the same attention all the way to the cellar.

The fermentation lasts 30 to 40 days depending on the characteristics of that vintage. Before bottling, Barolo Albe ages in Slavonian oak barrels for approximately 36 months.

TASTING NOTES

Sheer ruby red color. Aromas of red fruits and plums, flowers, pine needles and brushwood. Albe is a harmony of sophisticated and delicate flavors supported by an elegant acidity and ripe tannins.

Ettore Germano (48_1)

BAROLO DOCG 'CERRETTA'

VINEYARD AREA: 8 ACRES

VARIETAL COMPOSITION: 100% NEBBIOLO

DATE PLANTED: 1978-1995

VINES PER ACRE: 2000

EXPOSURE: SOUTH, SOUTH-EAST, AT AN ALTITUDE OF 1,150-1,225' ABOVE SEA LEVEL

SOIL: VERY CALCAREOUS (LIMESTONE-RICH)

YIELD: 3 TONS/ACRE

AVERAGE ANNUAL PRODUCTION: 18,000 BOTTLES

HARVEST: THE SECOND HALF OF OCTOBER

THE GRAPES ARE HAND-HARVESTED INTO SMALL BOXES, DESTEMMED AND CRUSHED, AND FERMENTED WITH THE SKINS FOR 25-30 DAYS, WITH DAILY PUMPOVERS TO EXTRACT COLOR AND TANNINS. AFTER DRAINING OFF THE WINE IS AGED FOR AT LEAST TWO YEARS IN FRENCH OAK BARRELS, MOSTLY OF 700 LITERS, 20% NEW AND THE REST FROM 2-4 YEARS OLD.. AFTER 15-18 MONTHS OF AGING IN THE BOTTLE THE WINE IS READY FOR RELEASE.

TASTING NOTES: INTENSE GARNET RED, WITH FAINT ORANGE GLINTS; THE AROMA IS OF RIPE FRUIT AND VANILLA, WHICH AFTER A FEW YEARS IN THE BOTTLE EVOLVE INTO LEATHER, TOBACCO, AND DARK CHOCOLATE, STILL REDOLENT OF BERRIES. THE WINE IS FULL ON THE PALATE, WITH SOLID TANNINS (ROUNDED OUT BY AGING IN SMALLER BARRELS) THAT LEAVE THE MOUTH CLEAN; THE FINISH IS LONG AND INTENSE.

SERVE WITH FLAVORFUL MAIN COURSES, GAME, OR AGED CHEESES. PERFECT AFTER DINNER AS A 'MEDITATION WINE' WITH NUTS AND CHEESE.

Ettore Germano (48_2)

BAROLO DOCG PRAPÒ

VINEYARD AREA: 2.5 ACRES (1 HECTAR) IN THE PRAPÒ VINEYARD IN SERRALUNGA D'ALBA

VARIETAL COMPOSITION: 100% NEBBIOLO

DATE PLANTED: 1967

VINE DENSITY: 1,520/ACRE (4200 / HECTAR)

EXPOSURE: SOUTH-EAST, AT AN ALTITUDE OF 1,150-1,225'FEET (360 METERS) ABOVE SEA LEVEL

SOIL: CALCAREOUS (LIMESTONE-RICH)

YIELD: 3 TONS/ACRE (7 TONS/HECTAR)

AVERAGE ANNUAL PRODUCTION: 5,000 BOTTLES

HARVEST: AROUND THE MIDDLE OF OCTOBER

THE VINIFICATION IS TRADITIONAL: THE GRAPES ARE HAND-PICKED INTO SMALL BOXES, DESTEMMED, CRUSHED, AND FERMENTED WITH THE SKINS FOR ABOUT 30 DAYS. AFTER 24 MONTHS IN LARGE OAK BARRELS (2,000L) AND 15 MONTHS OF AGING IN THE BOTTLE, THE WINE IS READY FOR RELEASE.

TASTING NOTES: GARNET RED WITH ORANGE GLINTS; INTENSE AROMA OF DRIED FRUITS, WITH HINTS OF LICORICE AND RED CURRANTS. IN THE MOUTH THE WINE IS WARM AND FULL-BODIED, THE ROUND TANNINS LEAVE THE MOUTH CLEAN AND WITH PERSISTENT RETRONASAL AROMAS. CLASSIC TRADITIONAL BAROLO, IN OTHER WORDS.

SERVING SUGGESTIONS: THE BEST ACCOMPANIMENTS WOULD BE VARIOUS TYPES OF MAIN COURSE, AGED CHEESES, OR DRIED FRUIT, FOR SIP PING WITH FRIENDS.

Palladino (49_1)

PALLADINO
SERRALUNGA D'ALBA

BAROLO D.O.C.G.
Ornato

VINEYARD: Nebbiolo 100%. Subvarietal: Lampia, Michet.

SOIL COMPOSITION: Calcareous clay.

EXPOSURE: South.

MICROCLIMATE: Fairly temperate.

TOTAL SURFACE AREA: 0,30 ha.

YEALD/Ha: 7.000 kg/ha.

COLOUR: Ruby red with purple and garnet reflections.

AROMA: Balsamic, reminiscence of the soil where the grapes grow.

TASTE: Full-bodied with a slightly bitter aftertaste, excellent structure and good predisposition to ageing.

FOOD AFFINITY: Braised or roasted meats, wild game, seasoned and strong cheeses. To be decanted and served at a temperature of 20° C.

Palladino (49_2)

BAROLO D.O.C.G.
Parafada

VINEYARD: Nebbiolo 100%. Subvarietal: Lampia, Michet.

SOIL COMPOSITION: Calcareous clay.

EXPOSURE: South.

MICROCLIMATE: Fairly temperate.

TOTAL SURFACE AREA: 0,67 ha.

YEALD/Ha: 7.000 kg/ha.

COLOUR: Ruby red with purple and garnet reflections.

AROMA: Slight scent of black raspberries and cherries with a background of vanilla. Ethereal with good structure.

TASTE: Full-bodied, austere and concentrated with a good balance of acidity and tannin.

FOOD AFFINITY: Braised or roasted meats, wild game, seasoned and strong cheeses. To be decanted and served at a temperature of 20° C.

Poderi e Cantine Oddero (50)

ODDERO

Barolo DOCG

Variety: 100% Nebbiolo

Vineyards

Location: La Morra, Santa Maria Bricco Chiesa, Plot 18, Sub-plots 503, 553, 443, 442, 552, 444, 483, 445, 612, 461
Characteristics: 200 meters above sea level, southeastern exposure, 2.7 x 0.9 meters spacing, 4,000 vines/hectare density, Guyot upward-trained vertical-trellised training system.
Age: 30 years - *Size:* 2.7 hectares - *Yield:* 70 quintals/hectare

Location: Castiglione Falletto Bricco Fiasco, Plot 7, Sub-plots 35, 36, 37, 38, 134
Characteristics: 250 meters above sea level, southwestern exposure, 2.7 x 0.9 meters spacing, 4,000 vines/hectare density, Guyot upward-trained vertical-trellised training system.
Age: 40 years - *Size:* 2.5 hectares - *Yield:* 70 quintals/hectare

Harvesting period: First 10 days of October

Production: About 30,000 bottles

Vinification: Rigorous manual selection of grapes in the vineyard. Fermentation and maceration in stainless steel for 20 days at a controlled temperature of 28-29 °C, followed by malolactic fermentation. The grapes from the two vineyards are kept separate throughout the process.

Maturing: The wines coming from the different vineyards are aged separately in 40, 60 and 75 –hectoliter Slavonian and Austrian oak barrels for 30 months. The wines are blended in the spring, then bottled at the end of the summer and aged in the bottle for another 6 months before release on the market.

Sensory characteristics: Brilliant garnet-red color, with warmer tinges developing over time. Unmistakeable bouquet, very pleasant, intense yet ethereal, very persistent, at times spiced, with aromas of truffles, liquorice and dried flowers. Dry taste, with important, full tannins.

Notes
What do you think of Hemingway? So asked readers of journalist Gianni Brera, who answered their literary and footballing questions in his column "L'accademia di Brera" ("Brera's academy") in the newspaper "La Repubblica". His reply, in "Sbornie prese per sport" ("Drunk for the sake of sport") on 12/12/1986: "Ernest drank Valpolicella because he was around Venice, and in those days the Oddero brothers of La Morra couldn't get their wines to him. But if he'd known Oddero Barolo he would not have turned to the liquor which ultimately killed him."

Conterno Fantino (55_1)

BAROLO "SORÌ GINESTRA" D.O.C.G.

Village: Monforte d'Alba
Cru: Ginestra
Varietal: Nebbiolo 100%
Vineyard planted in: 1971 - 1983 - 1988
Soil types: sand 27%, silt 51%, clay 22%
Exposure: South
Altitude: from 340 to 360 meters above sea level
Vineyard slope: 30%
Training system: Guyot
Density: 4.500/5.000 vines per ha
Harvest method: selected and picked by hand in mid-October
Average production: 5 t/ha - 35 hl/ha
Vinification: alcoholic fermentation and maceration on the skins in stainless steel horizontal fermenters for 8 - 15 days
Ageing: 24 months in French oak barrels, then in bottle for 12 months
First harvest: 1982
Bottles produced: approximately 12,500
Bottle sizes: 750 ml - 1500 ml - 3000 ml

This wine is made from organically grown grapes certified by CCPB srl - Bologna

Conterno Fantino (55_2)

BAROLO "VIGNA DEL GRIS" D.O.C.G.

Village: Monforte d'Alba
Cru: Ginestra
Varietal: Nebbiolo 100%
Vineyard planted in: 1978 - 2000
Soil types: sand 50%, silt 34%, clay 16%
Exposure: South, South-East
Altitude: from 300 to 320 meters above sea level
Vineyard slope: 30%
Training system: Guyot
Density: 4.500/5.000 vines per ha
Harvest method: selected and picked by hand in mid-October
Average production: 5 t/ha - 35 hl/ha
Vinification: alcoholic fermentation and maceration on the skins in stainless steel horizontal fermenters for 8 - 15 days
Ageing: 24 months in French oak barrels, then in bottle for 12 months
First harvest: 1982
Bottles produced: approximately 6,500
Bottle sizes: 750 ml - 1500 ml - 3000 ml

This wine is made from organically grown grapes certified by CCPB srl - Bologna

Conterno Fantino (55_3)

BAROLO "MOSCONI" D.O.C.G.

Village: Monforte d'Alba
Cru: Mosconi
Varietal: Nebbiolo 100%
Vineyard planted in: 1960 - 1999
Soil types: sand 26%, silt 50%, clay 24%
Exposure: South
Altitude: from 360 to 380 meters above sea level
Vineyard slope: 30%
Training system: Guyot
Density: 4.500/5.000 vines per ha
Harvest method: selected and picked by hand in mid-October
Average production: 5 t/ha - 35 hl/ha
Vinification: alcoholic fermentation and maceration on the skins in stainless steel horizontal fermenters for 8 - 15 days
Ageing: 24 months in French oak barrels, then in bottle for 12 months
First harvest: 2004
Bottles produced: approximately 6,000
Bottle sizes: 750 ml - 3000 ml

This wine is made from organically grown grapes certified by CCPB srl - Bologna

Conterno Giacomo (56)

CONTERNO

BAROLO
Denominazione di Origine Controllata e Garantita
"CASCINA FRANCIA"
2009

Grapes:	100% Nebbiolo
Vineyard planted:	1974
Production area:	"Cascina Francia" vineyard located in Serralunga d'Alba
Growing system:	Guyot
Altitude:	400 mt. above sea level
Exposure:	South-West
Soil characteristics:	Moderately dense limestone marl veined with sandstone rocks
Harvest time:	Around mid-October
Production:	around 25,000 bottles
Yield:	About 35 Q per ha.
Fermentation:	Long and traditional at about 30° C in wooden vats
Duration of fermentation:	around 3/4 weeks
Ageing:	4 years in large oak barrels
Alcohol:	around 14,5% Vol.
Total acidity:	about 5,6 gr/Lt
Sugar:	No sugar residue
Dry extract:	31 gr/Lt
Characteristics:	Colour tending to garnet red Intense scent with notes of tobacco, mint, spice and dried fruit. Voluminous and expansive on the palate
Suggestions:	It goes well with red meat, game and cheese

Poderi Luigi Einaudi (62_1)

Barolo
"Costa Grimaldi" docg

The result of an accurate selection of grapes from the Einaudi vineyard in Barolo. A great wine of the Langa, garnet red with a shade of amber, of intense and embracing fragrance, full bodied but smooth tannic taste, with a long spiced final taste. Derived from the Nebbiolo grapes from the Costa Grimaldi vineyard owned by the Einaudi estate, facing Barolo. The tannic characteristics guarantee a long life in bottle. With time it acquires complexity as the ethereal aromas of spices, truffles and leather.
The annual production is of 7.000 bottles.

Species of vine
Typology: Nebbiolo
Rootstock: 420 A
Bushiness: 4800 vines/ha
Plant breeding: Guyot

The Vineyards
Zona: Comune di Barolo (Cuneo)
Cru: Terlo - Costa Grimaldi 310 m.
Soil: marly-calcareous
Aspect: South-East
Area of the vineyard: 1,60 ha
Year of planting: 1977
Average yield per hectare: 60 q.

The wine making
Fermented in steel tanks under strict temperature control (30°C-32°C), post-fermentation maceration, racked after 20/22 days, decanted, then gotten induced malolactic treatment and completed always at a controlled temperature. Matured in barriques and oak casks totally for thirty months, then refinement in bottles for one year.

Poderi Luigi Einaudi (62_2)

PODERI LUIGI EINAUDI

Barolo
"nei Cannubi" docg

Born in the Cannubi vineyard of the Einaudi estate in the Cannubi area of Barolo at 220 m., positioned South South-East. The land is composed of grey white Sant'Agata marne, rich in sand. A wine of great class that expresses the elegance of the territory, of a brilliant garnet red turning slightly amber with time, exuberant in its fragrance of fruit and spices, of great body, full and velvety with a long final taste of goudron and spices. The tannic characteristics guarantee a very long life in bottle in the best vintages. The annual production is of 12.000 bottles.

Species of vine
Typology: Nebbiolo
Rootstock: 420 A / SO4
Bushiness: 4800-5000 vines/ha
Plant breeding: Guyot

The Vineyards
Zona: Comune di Barolo (Cuneo)
Cru: Cannubi 220 m.
Soil: Grey white marne of Sant'Agata (30% sand, 55% clay, 15% limestone)
Aspect: South/South-East
Area of the vineyard: 2,00 ha
Year of planting: 1962-1991-2002
Average yield per hectare: 60 q.

The wine making
Fermented in steel tanks under strict temperature control (30°C-32°C), post-fermentation maceration, racked after 20/22 days, decanted, and by maintaining the right temperature malolactic process is induced. Eighteen months of maturation in barriques, twelve months in big barrels, thereafter a long period in bottle to complete the aging.

索引

Giovanni Rosso (66_1)

I VINI

BAROLO D.O.C.G. "CERRETTA"

BAROLO
CERRETTA

Denomination:	Barolo D.O.C.G. CERRETTA
Grape Variety:	100% Nebbiolo
Alcohol:	14.0% vol
Production:	8,000 bottles
First year of production:	1996

Vineyard Characteristics

Soil, Aspect, Elevation:	Calcareous clay/marl, South-East facing, 360m
Locality:	2 miles (3km) north of Serralunga d'Alba
Training:	Guyot

Wine-Making

Grape-Harvest:	Harvested by hand
Fermentation:	Vinification in cement tanks with wild yeasts
	Fermentation over 30 - 35 days with Remontages, Delestages, and daily punching-down of the cap
Ageing:	Ageing takes place in large 25 hl Fontainebleau Forest French oak casks for 36 months
Bottling:	The wine is bottled without filtration

Description of Wine

Colour:	Concentrated, bright Garnet
Bouquet:	Violets, roses and cherry fruit with notes of coffee and layers of complexity
Taste:	The palate shows a wealth of Serralunga characteristics: full body, silky tannins, complexity and freshness. Bright red fruits combine with savoury, earthy flavours, a complete structure and a long finish
Bottle Formats:	375 ml, 750 ml, 1,5 Lt, 3 Lt, 5 Lt
Pairings:	Stews, red meats and darker game such as Wild Boar

Other Information

Total Acidity:	6.20 g / l
Sugar:	0 g / l
pH:	3.6
Serving Temperature:	16 °-17 °C
Glasses:	Burgundy balloons

Giovanni Rosso (66_2)

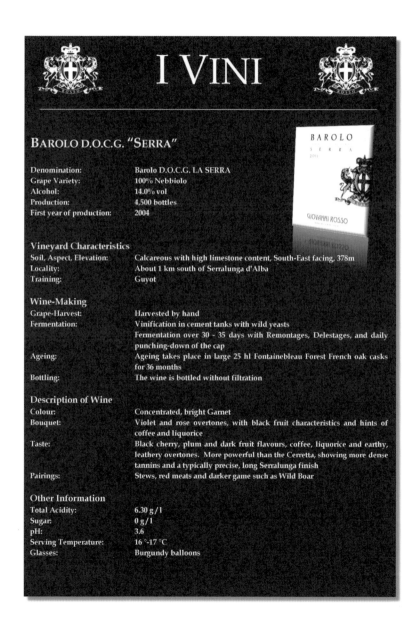

I VINI

BAROLO D.O.C.G. "SERRA"

Denomination:	Barolo D.O.C.G. LA SERRA
Grape Variety:	100% Nebbiolo
Alcohol:	14.0% vol
Production:	4,500 bottles
First year of production:	2004

Vineyard Characteristics

Soil, Aspect, Elevation:	Calcareous with high limestone content, South-East facing, 378m
Locality:	About 1 km south of Serralunga d'Alba
Training:	Guyot

Wine-Making

Grape-Harvest:	Harvested by hand
Fermentation:	Vinification in cement tanks with wild yeasts
	Fermentation over 30 - 35 days with Remontages, Delestages, and daily punching-down of the cap
Ageing:	Ageing takes place in large 25 hl Fontainebleau Forest French oak casks for 36 months
Bottling:	The wine is bottled without filtration

Description of Wine

Colour:	Concentrated, bright Garnet
Bouquet:	Violet and rose overtones, with black fruit characteristics and hints of coffee and liquorice
Taste:	Black cherry, plum and dark fruit flavours, coffee, liquorice and earthy, leathery overtones. More powerful than the Cerretta, showing more dense tannins and a typically precise, long Serralunga finish
Pairings:	Stews, red meats and darker game such as Wild Boar

Other Information

Total Acidity:	6.30 g / l
Sugar:	0 g / l
pH:	3.6
Serving Temperature:	16 °-17 °C
Glasses:	Burgundy balloons

Poderi Colla (69)

BAROLO DOCG
BUSSIA
Dardi le Rose – Bussia di Monforte

The Dardi di Bussia zone is historically considered one of the great Barolo crus. It is mentioned as one of the top locations, among other documents, in L. Fantini's 1880 publication *on Viticulture and Oenology in the Province of Cuneo* and in the *Atlas of the Langhe Area's Great Vineyards*, published by Arcigola Slow Food.

IN THE VINEYARD
Variety : 100% Nebbiolo
Exposure : southern, Southwesterly
Altitude : approximately 300 to 350 meters above sea level
Vines planted in : 1970 - 1985
Vines per hectare : approximately 4,000
Wine yield per hectare : around 4,000 liters
Harvest time : October 10th-20th

CELLAR
Grapes are hand-picked, destalked and crushed; the must successively macerates on the skins for 12/15 days at a maximum temperature of about 30° C., and malolactic fermentation ensues, completed before winter. In the spring, the wine sojourns in oak casks for a period from 24 to 28 months.

Alcohol :   13-14 % by vol.
Total acidity :   5-5.5 grams per liter of tartaric acid
Dry extract :   27-30 grams per liter

TASTING NOTE
Deep, intense ruby-red in colour, with a rich and refined bouquet, delicately spiced, bringing out floral tones, mingled with red berry fruit and tar. Full-bodied and strongly structured, it stays on the palate with supple tannin sensations, covered with fragrant notes of ripe fruit and liquorice. A traditional, "majestic" Nebbiolo with the might, balance and complexity of a symphony orchestra! Extraordinary ageing capacity.
Despite its strong personality,Barolo can be enjoyed frequently with red meat and ground game, even – why not? – with very rich fish dishes, such as baked turbot, or a dinner featuring speciality cheeses.

Serving temperature: 17-19°C – serve in a balloon glass – we advise decanting very old vintages before drinking to allow them to breathe.
.

Luigi Pira (70_1)

Barolo Vignarionda DOCG

VINEYARD	COMMUNE OF SERRALUNGA D'ALBA
VARIETY	NEBBIOLO 100%
YEAR OF PLANTING	1994
SURFACE	1 HA
SOIL	LIMESTONE AND CLAY
ALTITUDE	330 M. A.S.L.
EXPOSURE	SOUTH-WEST
YIELD	50-55 QT/HA
PRODUCTION IN BOTTLES	AROUND 4 THOUSAND

The wine-making process of **Barolo Vignarionda** takes place in horizontal fermentation tanks with temperature control for 12-15 days. The wine matures over 24 moths, partly in small wooden barrels and partly in large wooden barrels.

Intense garnet red color. The nose is complex, elegant, with hues that range from wild berries to sweet tobacco and to deep and enticing balsamic notes. In the mouth this wine is rich, balanced and with an extraordinary persistence.

Luigi Pira (70_2)

Barolo Margheria DOCG

VINEYARD	COMMUNE OF SERRALUNGA D'ALBA
VARIETY	NEBBIOLO 100%
YEAR OF PLANTING	1959
SURFACE	1.5 HA
SOIL	LIMESTONE AND CLAY
ALTITUDE	340 M. A.S.L.
EXPOSURE	SOUTH
YIELD	50-55 QT/HA
PRODUCTION IN BOTTLES	AROUND 7 THOUSAND

The wine-making process of **Barolo Margheria** takes place in horizontal fermentation tanks with temperature control for 12-15 days. The wine ages in large 25-hl wooden barrels for 24 months.

Garnet red color. Very intense and complex nose with notes of fruit and flowers and hints of leather and fresh hay. In the mouth it is full-bodied, elegant, with pleasant tannins that are typical of Nebbiolo and of the wines made in the area of Serralunga d'Alba. Perfect balance between tradition and innovation.

Luigi Pira (70_3)

Barolo Marenca DOCG

VINEYARD	COMMUNE OF SERRALUNGA D'ALBA
VARIETY	NEBBIOLO 100%
YEAR OF PLANTING	1990
SURFACE	2 HA
SOIL	LIMESTONE AND CLAY
ALTITUDE	350 M. A.S.L.
EXPOSURE	SOUTH/SOUTH-WEST
YIELD	50-55 QT/HA
PRODUCTION IN BOTTLES	AROUND 9 THOUSAND

The wine-making process of **Barolo Marenca** takes place in horizontal fermentation tanks with temperature control for 12-15 days. The wine matures over 24 moths, partly in small wooden barrels and partly in large wooden barrels.

Intense garnet red color. The nose is fresh, with complex perfumes that range from wild berries to vanilla, chocolate and aromatic herbs. On the palate it is very vigorous, rich in extracts, with lively and pleasant tannins that are rounded through a skilful use of wood. A powerful, tantalizing wine with a great ageing potential.

Attilio Ghisolfi (74_1)

BAROLO BUSSIA

Planting density:	over 5000 vines/ha.
Altitude:	300/350 m. asl.
Average age:	5 /13 years.
Position:	facing south, south-west
Soil:	marly, white tufa.
Vinification:	As with the other wines, only wild yeasts are used in the fermentation. The maceration on the skins lasts around 10 days, at a controlled temperature of between 25 and 30° C.
Maturing:	in oak casks holding 25-30 Hl for 30 months.
Ageing:	in the bottle for 6 months, at a controlled temperature of 15° C in the cellar.
Colour:	ruby, almost garnet red.
Nose:	flowery, rich, expansive and long.
Taste:	well-balanced, mature, deep and nicely tannic; good for drinking from an early age.
Storage temperature:	between 13° and 15° C.
Serving temperature:	16° C.
Pairings:	red meat roasts, game, truffle-flavoured dishes and mature cheeses.
Harvest:	2009
Bottles produced:	approx. 9.440
Awards:	2005 90 punti wine entusiasm 2005 3 stars decanter 2006 3 stars decanter 2007 93 punti wine spectator

Attilio Ghisolfi (74_2)

BAROLO BRICCO VISETTE

Average age: 20-50 years.

Planting density: over 5500 vines/ha

Altitude: 300/350 m. asl.

Position: facing south, south-west

Soil: marly, white tufa.

Vinification: maceration on the skins for 12 days, with careful controlling of the temperature, and natural fermentation brought on by native yeasts.

Maturing: 70% in oak casks holding 25-30 Hl, and 30% in small oak barriques for 30 months.

Ageing: in the bottle for 10 months, at a controlled temperature of 15° C.

Colour: dark garnet red.

Nose: elegant and intense red fruit, menthol overtones (when younger), spices.

Taste: rich and majestic, with medium tannic structure.

Storage temperature: between 13° and 15° C.

Serving temperature: 16° C.

Pairings: red meat roasts, game, truffle-flavoured dishes and mature cheeses.

Harvest: 2009

Bottles produced: approx. 9.857

Awards:
2000 94 punti wine spectator
2001 94 punti wine spectator
2001 tre bicchieri gambero rosso
2003 93 punti wine spectator
2005 91 punti wine entusiasm
2005 tre bicchieri gambero rosso
2006 4 stars decanter
2006 94 punti wine spectator

Giacomo Fenocchio (75_1)

Barolo docg Villero 2011

Location:
The historic "Villero" cru in the village of
Castiglione Falletto
Grape Variety: 100%Nebbiolo, varieries-
Michet,
Lampia and Rose
Vineyard Area: 1 Hectare
Exposure: Southwest
Altitude: Hilly zone at around 300 meters a.s.l.
Soil: Elveziano with clayey and calcareous
sediments, rich in iron.
Average age of the vines: 60 years
Grape yield per hectare: 65 Quintals
Harvest: In the first fifiteen days of Qctober.
Vinification: Traditonal method of long
maceration: natural fermentation without
odded yeasts for 40 days in stainless steel
tanks.

Aging: 6 months in stainless steel tanks,
30 months in large capacity Slavonian oak
casks (35-50hl) and a successive maturation
in the bottle.
Production: 5000 bottles a year
Tasting Notes: The color is a deep garnet red with ruby reflections, with a bright and
lively aspect. The scent is complex and rich with fruit, floral scents, spices, underbrush
and plums. It has an intense flavor with sweet tannis and a long aftertaste. The Barolo
Villero expresses itself with its characteristic stimulating scent accompanied by a dry.
but warm and elegant flavor which it retains for a long time.
Gastronomy: The Barolo Villero, with its structure and body characteristics, pairs well
withe dishes of meat, game and aged cheeses, typical dishes of the local cuisine, as well
as with many rich international dishes. The recommended serving temperature is 17-18
degress.

Alcohol level: Approx 14.5%vol
Total Acidity: Approx 5.37 g/L

Giacomo Fenocchio (75_2)

Barolo docg Villero 2011

Location:
Village of Monforte d'Alba-Bissia zone
Grape Variety: 100%Nebbiolo, varieries-
Michet and Lampia
Vineyard Area: 5 Hectare
Exposure: South-Southwest
Altitude: Hilly zone at around 300 meters a.s.l.
Soil: Elveziano with clayey and calcareous
sediments, rich in iron.
Average age of the vines: 30 years
Grape yield per hectare: 70 Quintals
Harvest: In the first fifiteen days of Qctober.
Vinification: Traditonal method of long
maceration: natural fermentation without
odded yeasts for 40 days in stainless steel
tanks.
Aging: 6 months in stainless steel tanks,
30 months in large capacity Slavonian oak
casks (35-50hl) and a successive maturation
in the bottle.
Production: 30,000 bottles a year

Tasting Notes: The color is a deep garnet red and the bouquet is fine and pleasting, with
intense scent of spiced rose ans licorice. The flavor is dry, warm, full bodied and
balance with pronounced tannins and a persistent aftertaste. A wine particularly
adapted for aging.
Gastronomy: The Barolo Bussia, with its structure and body characteristics, pairs well
withe dishes of meat, game and aged cheeses, typical dishes of the local cuisine, as well
as with many rich international dishes. The recommended serving temperature is 17-18
degress.

Alcohol level: Approx 15%vol
Total Acidity: Approx 5.13 g/L

Giacomo Fenocchio (75_3)

Barolo docg Castellero 2011

Location: *Barolo.*
Grape Variety: *100% Nebbiolo,*
variety- Michet and Lampia
Vineyard Area: *.0,6 Hectare*
Exposure: *West*
Altitude: *Hilly zone at around 300*
meters a.s.l.
Soil: *Helvetian, sandy with silt and*
clay.
Average age of the vines: *30 years*
Grape yield per hectare: *70 Quintals*
Harvest: *In the beginning of October*
Vinification: *Traditional method of*
long maceratio: natural fermentation
without added yeasts for 40 days in
stainless steel
tanks.
Aging: *6 months in stainless steel*
tanks,30 months in large capacity
Slavonian oak (casks 20/25 hl) and a
successive refinement in the bottle
Production: *3,000 bottles a year*
Tasting Notes: *The color is classic*
intense garnet red.
Clean and of good substance on the
nose with floral and fruity notes with
hints of liquorice and roots.
Wine of character with marked tannins . Velvety and harmonious
with notes of sweet spice and jam
Wine that can grow and evolve for many years
Gastronomy: *The Barolo Castelleroi is very good with red meat, game and aged*
cheese.Well as with many rich international dishes.
The recommended serving temperature is 17-
18 degrees,
Alcohol level: *Approx. 15% Vol*
Total Acidity: *Approx. 5.35 g/L*

Abborna Marziano (85_1)

BAROLO DOCG
Pressenda

GRAPE-VARIETY: Nebbiolo

VINEYARDS' LOCATION: Loc. Castelletto, Monforte d'Alba

HECTARES (ESTATE-OWNED): 3

VINEYARDS' AGE: 34 years

PLANTS PER HECTARE: 5.000

GRAPES' YIELD: 70 quintals per hectare

WINE-MAKING: fermentation takes place in rotating macerators

AGING: 12 months in tonneaux of 500L and 24 months in oaken barrels of 30 HL. Refining in bottle for 12 months

FIRST YEAR OF PRODUCTION: 1996

BOTTLES PRODUCED: 19.000

索引

BAROLO DOCG
Terlo Ravera

GRAPE-VARIETY: Nebbiolo

VINEYARDS' LOCATION: Loc. Ravera, Novello

HECTARES (ESTATE-OWNED): 4

VINEYARDS' AGE: 33 years

PLANTS PER HECTARE: 5.000

GRAPES' YIELD: 70 quintals per hectare

WINE-MAKING: fermentation takes place in rotating macerators

AGING: 12 months in tonneaux of 500L and 24 months in oaken barrels of 30 HL. Refining in bottle for 12 months

FIRST YEAR OF PRODUCTION: 1990

BOTTLES PRODUCED: 28.000

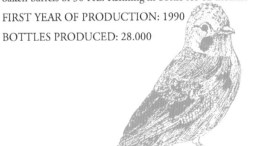

Abborna Marziano (85_3)

BAROLO DOCG
Cerviano

VITIGNO: Nebbiolo

ZONA DI PRODUZIONE: Loc. Cerviano, Novello

SUPERFICIE IN PROPRIETÀ: Ha 1.5

ETÀ DEI VIGNETI: 18 anni

DENSITÀ DI IMPIANTO: 5.000 piante per ettaro

RESA D'UVA: 65 quintali per ettaro

VINIFICAZIONE: in vasche di acciaio a 32°C in condizioni di temperatura controllata. Ha luogo un periodo di 7-8 giorni di macerazione sulle bucce. Subito dopo la fermentazione malolattica, il vino viene travasato

INVECCHIAMENTO: il vino permane per 36 mesi in botti grandi in rovere di Slavonia.
Affinamento in bottiglia per 12 mesi

PRIMO ANNO DI PRODUZIONE: 2001

BOTTIGLIE PRODOTTE: 6.500
Questo vino viene prodotto solo nelle migliori annate

索引

Teo Costa (88_1)

BAROLO
Denominazione di Origine Controllata e Garantita

Tipo di vino: rosso invecchiato di enorme struttura e potenza

Collocazione del vigneto: Comune di La Morra, località Sorello

Giacitura e tipo di tereno: collina esposta ed Ovest particolarmente ricca di silice e di calcare con una spiccata stratificazione di tufo e pietra e percentuali di argilla minoritarie

Varietà di uva: Nebbiolo Lampia 100%

Cure nella preparazione: selezione delle uve con vinificazione in botti di rovere, macerazione e frequenti follature delle vinacce

Tenore Alcool: 14% - **Acidità:** 5,6 gr/l

Invecchiamento: 30 mesi in botte di rovere grande in bottiglia: 6 mesi

Tipo di bottiglia: antica anni 50

Imballo: casse da n. 6 bottiglie coricate

Disponibilità: quarto anno successivo alla vendemmia

Colore: rosso granato profondo con riflessi aranciati

Profumo: intenso, nobile con grande persistenza

Sapore: di grande armonia e ed aristocrazia. Eccezionalmente pieno, generoso e vellutato

Epoca di consumo: raggiunge l'apice qualitativo dopo 5-6 anni e si evolve positivamente fino a 18-20 anni

Temperatura di servizio: 16° - 17° C

Teo Costa (88_2)

MONROJ®
BAROLO
Controlled and guaranteed origin

Although for decades wine-makers in Piemonte have somehow "rested on their laurels" and Barolo and Barbaresco have slowly suffered a loss in popularity, Barolo remains - being fully acknowledged today as - one of the greatest wines in the world, rich in tradition and history, incomparable in character. The growing area is strictly limited by the D.O.C.G. regulation like for the Barbaresco wine and includes only 11 villages. Barolo and La Morra are part of the so-called morphological "tortorian" sub-zone where the composition of the soil, the exposure and the climate give the wine mellow and fruity sensations along with strength and structure. From the grapes of its "Monroj" vineyard at La Morra, beautifully placed on a hill top facing the Monviso mountain peak and the French Alps, the Teo Costa Winery produces its "Monroj", a Barolo of rare complexity and delicacy.

Type of wine: aged red wine with firm structure and superb concentration.

Vineyard location: town of La Morra, location Sorello.

Soil and exposure: hilly soil, particularly rich in silica and limestone and characterized by tufa and stone stratification with little clay. Western exposure.

Grape variety: 100% Nebbiolo Lampia.

Vinification: accurate selection of grapes. Vinification in oak casks with short-termed maceration of skins - with extraction of distinctive aromas and tannins - and constant stirring of the must.

Alcohol content: 14.5 % - Acidity 5.6 grams per litre.

Aging: 30 months in big oak barrels. 6 months in bottles.

Type of bottle: old fashioned bottle like in the 50's. Packaging: 6-bottle wooden cases (bottles in horizontal position).

Release: in the 4th year after harvesting.

Colour: deep garnet red with orange reflections.

Smell: noble, intense, and very persistent.

Taste: very rich and harmonious, extraordinarily full, generous and velvety.

Evolution: this wine is usually at its best after 5-6 years and keeps its excellent quality for 18-20 years.

Serving temperature: 18 -19 °C.

索引

Bric Cenciurio_020
Via Roma 24
12060 Barolo (CN)
Tel：0173 56317

Bruna Grimaldi_005
Via Parea 7
12060 Grinzane Cavour (CN)
Tel：0173 262094

Bussia Soprana_045
Località Bussia Soprana 81
12065 Monforte d'Alba (CN)
Tel：039 305182

Cadia_035
Strada Roddi-Verduno 62
12060 Roddi (CN)
Tel：0173 615398

Ca'Rome'_018
Strada Rabajà 86
12050 Barbaresco(CN)
Tel：0173 635126

Cascina Adelaide_004
Via Aie Sottane 14
12060 Barolo (CN)
Tel：0173 560503

Dosio Vigneti_027 p.171
Loc. Serradenari 6
12064 La Morra (CN)
Tel：0173 50677

Ettore Germano_048 p.203
Borgata Cerretta, 1
12050 Serralunga d'Alba (CN)
Tel：0173 613528

Fratelli Casetta_026 p.170
Via Castellero N, 5
12040 Vezza d'Alba (CN)
Tel：0173 65010

Fratelli Giacosa_040 p.192
Via 20 Settembre, 64
12052 Neive (CN)
Tel：0173 67013

Gemma_032 p.179
Via Alba 11/C
12050 Serralunga d'Alba (CN)
Tel ：0173 262360

Giacomo Fenocchio_075 p.222
Loc. Bussia, Sottana 66
12065 Monforte d'Alba (CN)
Tel：0173 78675

索引

索引

Vietto_033 p.180

Loc. Panerole, 22
12060 Novello (CN)
Tel：0173 731379

Virna_021 p.161

Via Alba 73
12060 Barolo (CN)
Tel：0173 56120

國家圖書館出版品預行編目 (CIP) 資料

Barolo Library 餐桌上的義大利酒王：
我的 100 瓶巴洛羅紅酒評審指南 / 黃筱雯著
初版 . 臺北市：聚樂錄義大利美食有限公司出版 , 2015.07
　　面 ；　公分 . -- (美食旅人 ； 2)
ISBN 978-986-92022-0-6 （精裝）
1. 紅酒 2. 品酒 3. 義大利

463.814　　　　　104011932

美食旅人 002

Barolo Library
餐桌上的義大利酒王
我的 100 瓶巴洛羅紅酒評審指南

作　　　者 A u t h o r	黃筱雯 Xiaowen Huang
攝　　　影 Photographor	Anthony Gaudun/ Julien Nguyen-Kim
責 任 編 輯 E d i t o r	黃筱雯 Xiaowen Huang
美 術 設 計 Art Design	乙曼資訊科技有限公司 Yiman Infotek Co., LTD.
美 術 助 理 Art Assistant	林欣諭 Shinyu Lin
出 版 者 P u b l i s h e r	聚樂錄義大利美食有限公司 CLUBalogue Academy
地　　　址 A d d r e s s	104 臺北市中山區八德路二段 72 號 2 樓 Via 2 Giugno 11, 24065 Lovere BG, Italy
電　　　話 T e l e p h o n e	+886(0)2-2564-3890 +39 035 983 400
網　　　址 W e b s i t e	www.clubalogue.com
信　　　箱 E m a i l	info@clubalogue.com
匯 款 帳 號 銀　　　行 戶　　　名 帳　　　號	台新銀行 聚樂錄義大利美食有限公司 2089-01-0000-2646
製 版 印 刷	英倫國際文化事業股份有限公司
定　　　價	新台幣 650 元 / EUR 25
初 版 一 刷	2015 年 7 月 / July 2015

代理經銷 / 白象文化事業有限公司
402 台中市南區美村路二段 392 號
電話：(04)2265-2939
傳真：(04)2265-1171